Applied Mathematical Sciences
Volume 91

Applied Mathematical Sciences

(continued following index)

Brian Straughan

The Energy Method, Stability, and Nonlinear Convection

With 30 Illustrations

Springer-Verlag

New York Berlin Heidelberg London Paris
Tokyo Hong Kong Barcelona Budapest

Brian Straughan
Department of Mathematics
University of Glasgow
University Gardens
Glasgow, G12 8QW
United Kingdom

Editors

F. John
Courant Institute of
  Mathematical Sciences
New York University
New York, NY 10012
USA

J.E. Marsden
Department of
  Mathematics
University of California
Berkeley, CA 94720
USA

L. Sirovich
Division of
  Applied Mathematics
Brown University
Providence, RI 02912
USA

Mathematics Subject Classification: 76E15, 76E20, 76E30

Library of Congress Cataloging-in-Publication Data
Straughan, B. (Brian)
    The energy method, stability, and nonlinear convection/Brian
Straughan.
      p.   cm. — (Applied mathematical sciences ; 91)
    Includes bibliographical references and index.
    ISBN 0-387-97658-2 (Springer-Verlag New York Berlin Heidelberg). -
    ISBN 3-540-97658-2 (Springer-Verlag Berlin Heidelberg New York)
      1. Fluid dynamics.   2. Differential equations, Nonlinear-
-Numerical solutions.   3. Heat — Convection — Mathematical models.
    I. Title.   II. Series: Applied mathematical sciences (Springer-
Verlag New York Inc.);v. 91.
    QA1.A647 vol. 91
    [QA911]
    510 — dc20
    [532′.05′01515355]   91-22901

Printed on acid-free paper.

Photocomposed pages prepared from the author's T$_E$X files.
Printed and bound by R.R. Donnelley & Sons, Harrisonburg, VA.
Printed in the United States of America.

9 8 7 6 5 4 3 2 1

ISBN 0-387-97658-2 Springer-Verlag New York Berlin Heidelberg
ISBN 3-540-97658-2 Springer-Verlag Berlin Heidelberg New York

To

Luan & Kirsten

Preface

The writing of this book was begun during the academic year 1984–1985 while I was a visiting Associate Professor at the University of Wyoming. I am extremely grateful to the people there for their help, in particular to Dick Ewing, Jack George and Robert Gunn, and to Ken Gross, who is now at the University of Vermont.

A major part of the first draft of this book was written while I was a visiting Professor at the University of South Carolina during the academic year 1988–1989. I am indebted to the people there for their help, in one way or another, particularly to Ron DeVore, Steve Dilworth, Bob Sharpley, Dave Walker, and especially to the chairman of the Mathematics Department at the University of South Carolina, Colin Bennett.

I also wish to express my sincere gratitude to Ray Ogden and Professor I.N. Sneddon, F.R.S., both of Glasgow University, for their help over a number of years. I also wish to record my thanks to Ron Hills and Paul Roberts, F.R.S., for giving me a copy of their paper on the Boussinesq approximation prior to publication and for allowing me to describe their work here. I should like to thank my Ph.D. student Geoff McKay for spotting several errors and misprints in an early draft. Finally, I am very grateful to an anonymous reviewer for several pertinent suggestions regarding the energy-Casimir method.

Brian Straughan
Glasgow

Contents

9. Convection in Generalized Fluids   145

10. Time Dependent Basic States   154

11. Electrohydrodynamic and Magnetohydrodynamic Convection   161

12. Ferrohydrodynamic Convection   176

13. Convective Instabilities for Reacting Viscous Fluids Far from Equilibrium   190

1
Introduction

This book is primarily a presentation of nonlinear energy stability results obtained in convection problems by means of an integral inequality technique we refer to as the energy method. While its use was originally based on the kinetic energy of the fluid motion, subsequent work has, for a variety of reasons, introduced variations of the classical energy. The new functionals have much in common with the Lyapunov method in partial differential equations and standard terminology in the literature would now appear to be *generalized energy methods*. In this book we shall describe many of the new generalizations and explain why such a generalization was deemed necessary. We shall also explain the physical relevance of the problem and indicate the usefulness of an energy technique in this context.

There is another technique in continuum mechanics also sometimes referred to as the energy method. This method seeks to find the configuration of minimum potential energy and then attempts to establish it as the stable configuration. These ideas go back at least as far as the work of Kelvin (1887) on a perfect fluid and have been extended by Arnold (1965a,b,1966a,b). A comprehensive and authoritative review of the Arnold method up to 1985 is given by Holm et al. (1985): this paper covers many interesting topics in fluid mechanics. A later connected work developing and extending this technique to an idea known as the energy-momentum method is that of Simo et al. (1990). We do not describe these ideas and concentrate on the dynamic theory of Energy Stability whose modern revival has been inspired by the work of Serrin (1959a,b) and by Joseph (1965,1966).

A brief outline of the contents is now given. In chapter 2 we develop the ideas of the energy method on some simple one-space dimensional equations. Chapter 3 then extends the energy method to the equations for a linear viscous, incompressible fluid, and to the standard Bénard problem for a linear viscous, incompressible, heat-conducting fluid. In chapters 4 to 10 we systematically develop the energy method in a variety of contexts: half-space problems, geophysical problems, convection driven by surface tension, convection in other classes of fluid, time-dependent convection problems, and we study the connection with the Lyapunov method in partial differential equations. Chapters 11 to 13 adopt a different line of attack. They concentrate on technologically relevant and relatively new theories: dielectric fluids, ferrohydrodynamics, and chemically reacting mixtures. Where appropriate, a thermodynamic derivation of the necessary equations is given. Very little on energy stability theory has been developed on these theories and could profitably be done so in future. The main body of the book closes with a brief concluding chapter, number 14. Two appendices are included. The first collects some inequalities frequently used in energy stability theory. The second is a brief account of numerical solutions to the type of eigenvalue problem one encounters in linear and in nonlinear energy stability theory. Particular attention is paid to the compound matrix method.

We shall employ standard notation throughout the book. In chapter 2 we denote partial derivatives in the usual way; e.g., $\frac{\partial u}{\partial t}$ or by subscripts, e.g.,

$$u_t \equiv \frac{\partial u}{\partial t}, \qquad u_{xx} \equiv \frac{\partial^2 u}{\partial x^2}.$$

In chapters 3 to 13 standard vector and indicial notation is used as necessary.

While the chapters do follow on from one another in a logical manner, we have attempted to make each chapter as self-contained as possible. Thus, for example, we may specifically refer to the notation used.

2

Illustration of the Energy Method on Simple Examples and Discussion of Linear Theory

2.1 The Diffusion Equation

Before considering the nonlinear partial differential equations encountered in convection we approach the problem of stability or instability for the diffusion equation.

Let u be a solution to the diffusion equation

$$u_t = u_{xx}, \tag{2.1}$$

where for now, $-\infty < x < \infty$, $t > 0$, and suppose we wish to investigate the behaviour of u subject to initial data

$$u(x, 0) = u_0(x), \qquad -\infty < x < \infty. \tag{2.2}$$

The zero solution $u \equiv 0$ is a solution to (2.1), and it is the stability of this solution we wish to investigate. The key to the procedure is that for a solution (here $u \equiv 0$) to be stable it must be stable against *any* disturbance to which it may be subjected. In fact, we shall understand stability to be in the sense of asymptotic stability so that for the zero solution to (2.1) to be stable we mean all perturbations decay to zero as time progresses. From a practical viewpoint it is necessary that disturbances are damped out sufficiently rapidly. In this respect the energy method is very useful as it usually guarantees exponential decay.

To demonstrate instability of the steady solution $u \equiv 0$, on the other hand, it is sufficient to find *one* disturbance that either progressively grows in amplitude or at least remains bounded away from zero in the case of instability of the zero solution.

With the above in mind, since (2.1) is linear we consider a perturbation to the zero solution to (2.1) of the form

$$u(x, t) = e^{\sigma t} e^{ikx}, \tag{2.3}$$

where k is any real number. Despite the special appearance of this perturbation, we are able to study the instability of the zero solution to (2.1) against any periodic (in x) disturbance. The idea behind this is that we suppose u is periodic in x, but allow any periodic behaviour, then $u(x,t)$ may be written as

$$u(x,t) = \sum_{n=0}^{\infty} A_n e^{\sigma_n t} e^{ik_n x},   (2.4)$$

i.e., a Fourier series. The term in (2.3) is referred to as a Fourier mode. As it is sufficient for only one destabilizing disturbance to cause instability we need consider only (2.3), because by varying over all real numbers k we hence would pick up the most destabilizing term in (2.4).

For a representation of the form (2.3), equation (2.1) reduces to

$$\sigma = -k^2.$$

Since $k \in \mathbf{R}$, $\sigma < 0$ and so there is no unstable mode and all solutions of form (2.3) decay. Therefore, for the periodic disturbance situation considered above the zero solution to (2.1) is always stable.

Of course, the above analysis is for an infinite spatial region whereas many practical convection studies usually require a finite region of disturbance. Hence, without loss of generality, suppose now (2.1) holds on the spatial region $(0, 1)$ with boundary conditions

$$u(0, t) = u(1, t) = 0.   (2.5)$$

The only difference with the infinite region case for (2.1) is that k can no longer take all values in $\mathbf{R}$, for the class of periodic functions in x must all vanish at $x = 0, 1$. Hence, here we have a subclass of (2.3). As all solutions of form (2.3) are stable, so are all solutions of this subclass.

Even though we know $u \equiv 0$ is a stable solution to (2.1), (2.5) we shall obtain this result directly using an energy method, choosing an extremely simple example to illustrate the technique.

Let now u be a solution to (2.1), (2.5) that satisfies arbitrary initial data $u_0(x)$. We define an energy $E(t)$ by

$$E(t) = \frac{1}{2}\|u(t)\|^2,   (2.6)$$

where $\|\cdot\|$ denotes the norm on $L^2(0, 1)$, i.e.,

$$\|f\|^2 = \int_0^1 f^2\, dx.$$

Differentiate $E(t)$ and use (2.1),

$$\frac{dE}{dt} = \frac{1}{2}\int_0^1 \frac{\partial}{\partial t}u^2\,dx$$

$$= \int_0^1 uu_t\,dx$$

$$= \int_0^1 uu_{xx}\,dx.$$

Next, integrate by parts,

$$\int_0^1 uu_{xx}\,dx = -\int_0^1 u_x^2\,dx + uu_x\Big|_0^1$$

$$= -\|u_x\|^2.$$

Therefore, the energy equation becomes

$$\frac{dE}{dt} + \|u_x\|^2 = 0. \tag{2.7}$$

Next, use the Poincaré inequality

$$\|u_x\|^2 \geq \pi^2\|u\|^2,$$

for u with $u = 0$ at $x = 0, 1$. Using this in (2.7) gives

$$\frac{d}{dt}\|u\|^2 + 2\pi^2\|u\|^2 \leq 0$$

or, equivalently,

$$\frac{d}{dt}\left(e^{2\pi^2 t}\|u\|^2\right) \leq 0,$$

which leads to

$$\|u(t)\|^2 \leq e^{-2\pi^2 t}\|u_0\|^2. \tag{2.8}$$

Hence, $\|u(t)\| \to 0$ at least exponentially and the zero solution to (2.1), (2.5) is stable.

2.2 The Linear Diffusion Equation with a Linear Source Term

The problem is to study the stability of the zero solution to the equation

$$u_t = u_{xx} + au, \tag{2.9}$$

where a is a positive constant, with given initial data

$$u(x,0) = u_0(x). \tag{2.10}$$

When $a = 0$ it has been shown in §2.1 that the zero solution is stable always, regardless of whether the spatial domain is finite or infinite. Here things are different.

Spatial Region $x \in \mathbf{R}$. We use a *normal mode* analysis, and as before substitute

$$u(x,t) = e^{\sigma t + ikx}.$$

Then, from (2.9) we obtain

$$\sigma = -k^2 + a. \tag{2.11}$$

Now, $x \in \mathbf{R}$, therefore we have to analyse stability among *all* periodic disturbances, i.e., $k \in \mathbf{R}$. From (2.11) since $\sigma > 0 \Rightarrow$ instability, we conclude

$$k^2 < a \quad \Rightarrow \quad \text{instability},$$

i.e., any mode for which $k^2 < a$ is sufficient to yield an unstable solution. Therefore, if $a > 0$, there is always some $k^2 < a$ and so $a > 0 \Rightarrow$ instability *always*.

So, with the infinite spatial region, $x \in \mathbf{R}$, we have $u \equiv 0$ always stable, for u satisfying

$$u_t = u_{xx},$$

but if

$$u_t = u_{xx} + au,$$

then $u \equiv 0$ is always unstable no matter how small a $(a > 0)$.

Finite Spatial Region. Suppose now $x \in (0,1)$, u satisfies (2.9), and $u(0,t) = u(1,t) = 0$, $\forall t > 0$. Here (2.11) is still valid, but now $u = 0$ whenever $x = 0, 1$; i.e., the solutions may be thought of as being spatially periodic over $\mathbf{R}$ but *must* vanish at $x = 0, 1$, and at every multiple of 1.

For example,

$$u(x,t) = e^{\sigma t}\sin kx,$$

and then

$$u = 0 \quad \text{at} \quad x = 0, 1 \quad \Rightarrow \quad k = n\pi, \quad n = \pm 1, \pm 2, \ldots.$$

The Fourier-series is here a half-range series because cosine terms do not satisfy the boundary conditions. Therefore,

$$\sigma = -k^2 + a, \quad \text{where} \quad k^2 = n^2\pi^2, \quad n = \pm 1, \pm 2, \ldots.$$

For $\sigma > 0$, $k^2 < a$, but $k_{\min}^2 = \pi^2$. Hence, provided $a > \pi^2$, there is *always* instability, i.e., the mode $e^{\sigma t}\sin \pi x$ will grow. For $a < \pi^2$, all Fourier modes decay, and so there is linear stability. So $a = \pi^2$ is the stability-instability boundary. (It should be observed that the size of the region is crucial; for $x \in (0, L)$ we find a different stability boundary.)

Energy Stability for a Solution to (2.9). We apply the energy method to study the stability of the solution to the boundary-initial value problem,

$$\frac{\partial u}{\partial t} = \frac{\partial^2 u}{\partial x^2} + au, \qquad x \in (0,1), \ t > 0,$$
$$u(0,t) = u(1,t) = 0, \qquad \forall\, t \geq 0, \tag{2.12}$$
$$u(x,0) = u_0(x).$$

Multiply the differential equation $(2.12)_1$ by u and integrate over $(0,1)$ to find

$$\int_0^1 u\frac{\partial u}{\partial t}\,dx = \int_0^1 u\frac{\partial^2 u}{\partial x^2}\,dx + a\int_0^1 u^2\,dx.$$

Hence, with

$$E(t) = \frac{1}{2}\|u(t)\|^2 \quad \left(= \frac{1}{2}\int_0^1 u^2(x,t)\,dx\right),$$

this equation becomes (after integration by parts as before)

$$\frac{dE}{dt} = -\|u_x\|^2 + a\|u\|^2,$$
$$= -a\|u_x\|^2\left(\frac{1}{a} - \frac{\|u\|^2}{\|u_x\|^2}\right), \qquad (\|u_x\|^2 \neq 0),$$
$$\leq -a\|u_x\|^2\left(\frac{1}{a} - \max_{\mathcal{H}}\frac{\|u\|^2}{\|u_x\|^2}\right), \tag{2.13}$$

where $\mathcal{H}$ is the space of admissible functions over which we seek a maximum. Set

$$\mathcal{H} = \{u \in C^2(0,1) \mid u = 0 \text{ when } x = 0,1\}.$$

Now define R_E by

$$\frac{1}{R_E} = \max_{\mathcal{H}}\frac{\|u\|^2}{\|u_x\|^2};$$

then the energy inequality (2.13) may be rewritten

$$\frac{dE}{dt} \leq -a\|u_x\|^2\left(\frac{1}{a} - \frac{1}{R_E}\right).$$

If $a < R_E$, then $1/a - 1/R_E > 0$, say $1/a - 1/R_E = c\,(> 0)$, and so

$$\frac{dE}{dt} \leq -ac\|u_x\|^2.$$

Again using the Poincaré inequality, $\|u_x\|^2 \geq \pi^2\|u\|^2$, from (2.13) we deduce

$$\frac{dE}{dt} \leq -ac\pi^2\|u\|^2 = -2\pi^2 acE,$$

from which it follows that

$$\frac{d}{dt}(e^{2\pi^2 act}E) \leq 0.$$

This may be integrated to see that

$$E(t) \leq e^{-2\pi^2 act}E(0). \tag{2.14}$$

We have thus shown that if $a < R_E$,

$$E(t) = \frac{1}{2}\|u(t)\|^2 \to 0 \quad \text{as} \quad t \to \infty,$$

with the decay at least exponential in time.

The problem remains to find R_E. Recall that

$$R_E^{-1} = \max_{\mathcal{H}} \frac{\|u\|^2}{\|u_x\|^2}.$$

Let $I_1 = \|u\|^2$, $I_2 = \|u_x\|^2$. The Euler-Lagrange equations are found from

$$\frac{d}{d\epsilon}\frac{I_1(u+\epsilon\eta)}{I_2(u_x+\epsilon\eta_x)}\Big|_{\epsilon=0} = \delta\left(\frac{I_1}{I_2}\right) = \frac{I_2\delta I_1 - I_1\delta I_2}{I_2^2}$$

$$= \frac{1}{I_2}\left(\delta I_1 - \frac{I_1}{I_2}\Big|_{\max}\delta I_2\right)$$

$$= \frac{1}{I_2}\left\{\delta I_1 - \frac{1}{R_E}\delta I_2\right\}.$$

(Since δ refers to the "derivative" evaluated at $\epsilon = 0$, I_1/I_2 is here understood to be at the stationary value.) Therefore,

$$\delta I_1 - \frac{1}{R_E}\delta I_2 = 0. \tag{2.15}$$

Here

$$\delta I_1 = \frac{d}{d\epsilon}\int_0^1 (u+\epsilon\eta)^2 dx\Big|_{\epsilon=0},$$

where η is an arbitrary $C^2(0,1)$ function with $\eta(0) = \eta(1) = 0$, and

$$\delta I_2 = \frac{d}{d\epsilon}\int_0^1 (u_x+\epsilon\eta_x)^2 dx\Big|_{\epsilon=0}.$$

So (2.15) leads to

$$\int_0^1 (u\eta - R_E^{-1}\eta_x u_x)dx = 0.$$

Integration by parts shows that

$$\int_0^1 \eta(u_{xx} + R_E u)dx = 0.$$

Since η is arbitrary apart from the continuity and boundary condition requirements, we must have

$$\frac{d^2u}{dx^2} + R_E u = 0, \qquad u(0) = u(1) = 0. \tag{2.16}$$

Equation (2.16) is the Euler equation that gives an eigenvalue problem for R_E.

The general solution to (2.16) is

$$u = A \sin R_E^{1/2} x + B \cos R_E^{1/2} x,$$

where A, B are constants to be determined. The boundary condition $u(0) = 0$ shows $B = 0$, and so

$$u = \sin R_E^{1/2} x,$$

where we have taken $A = 1$, since we are primarily interested in R_E not u. The condition $u(1) = 0$ then shows that

$$\sqrt{R_E} = n\pi, \qquad n = \pm 1, \pm 2, \ldots .$$

This gives an infinite sequence of values for R_E (corresponding to stationary values of the quotient $\|u\|^2/\|u_x\|^2$),

$$R_E = \pi^2, 4\pi^2, 9\pi^2, \ldots .$$

For stability, we need $a < R_E(\min)$, and so $R_E = \pi^2$. In particular, therefore,

$$a < \pi^2$$

yields stability of the zero solution to (2.9), (2.10). This criterion has been derived by using energy inequalities. It is the same as the one found by the "normal mode" method. This is partly because the basic differential equation is linear. We are primarily interested in nonlinear differential equations. In this situation the energy method is very useful since it does yield precise practical information regarding the *nonlinear* stability of a basic solution (the solution $u \equiv 0$ in the examples above).

2.3 Nonlinear, Conditional Energy Stability for a Nonlinearly Forced Diffusion Equation

In order to investigate the effect of a nonlinear term on the stability of a solution to a partial differential equation (pde), we shall consider the boundary-initial value problem for the diffusion equation (2.1) but now with a quadratic nonlinear term forcing the right-hand side and also with a convective nonlinear term uu_x added to the left (such a convective non-linearity is typical in hydrodynamic stability questions).

We examine the stability of the zero solution to the boundary-initial value problem,

$$\frac{\partial u}{\partial t} + u\frac{\partial u}{\partial x} = \frac{\partial^2 u}{\partial x^2} + \beta u^2, \qquad x \in (0,1),\ t > 0,$$
$$u(0,t) = u(1,t) = 0, \qquad \forall\, t \geq 0, \tag{2.17}$$
$$u(x,0) = u_0(x),$$

where β is a positive constant.

If we attempt a linear analysis, that is linearize about the solution $u \equiv 0$, then since any perturbation u is assumed such that $|u|, |u_x| \ll 1$, u^2 and uu_x may be neglected. We are then left with the linear stability analysis of (2.1). Therefore, the zero solution is always *linearly stable*. The nonlinear terms, however, cannot be neglected. The effect of the u^2 term is to destabilize as we shall now show. The uu_x term in certain respects acts to stabilize and this is shown in §2.4.

To see the effect of the nonlinear term u^2 we multiply the differential equation $(2.17)_1$ by u and integrate over (0,1) to obtain

$$\frac{1}{2}\frac{d}{dt}\|u\|^2 = \int_0^1 u\frac{\partial^2 u}{\partial x^2}\,dx + \beta\int_0^1 u^3\,dx.$$

Note that the convective term integrates to zero,

$$\int_0^1 u^2\frac{\partial u}{\partial x}\,dx = \frac{1}{3}\left[u^3(1) - u^3(0)\right] = 0.$$

This is a key point as this feature carries over to fluid mechanics problems and allows us to remove otherwise troublesome terms.

Again, as before, integrating by parts,

$$\int_0^1 u\frac{\partial^2 u}{\partial x^2}\,dx = -\|u_x\|^2.$$

So, the energy equation in this case becomes

$$\frac{1}{2}\frac{d}{dt}\|u\|^2 = -\|u_x\|^2 + \beta\int_0^1 u^3\,dx. \tag{2.18}$$

Since $u(x, t)$ may be positive or negative we do not know a priori if $\int u^3 \, dx$ is one signed. In general, this term will lead to an instability. Therefore, may we recover any stability? Motivated by linear theory, which shows that if $|u| \ll 1$ the solution is always stable, we might guess that we shall have stability provided the initial data $u_0(x)$ is small enough.

We write

$$\int_0^1 u^3 \, dx = \int_0^1 u^2 u \, dx \leq \left(\int_0^1 u^4 \, dx \right)^{1/2} \left(\int_0^1 u^2 \, dx \right)^{1/2},$$

by use of the Cauchy-Schwarz inequality. From the Sobolev embedding inequality we know that (Appendix 1)

$$\int_0^1 u^4 \, dx \leq \frac{1}{4} \left(\int_0^1 u_x^2 \, dx \right)^2.$$

Using this leads to

$$\int_0^1 u^3 \, dx \leq \frac{1}{2} \left(\int_0^1 u_x^2 \, dx \right) \left(\int_0^1 u^2 \, dx \right)^{1/2}$$

$$= \frac{1}{2} \|u\| \, \|u_x\|^2. \tag{2.19}$$

Put (2.19) into (2.18) to find

$$\frac{1}{2} \frac{d}{dt} \|u\|^2 \leq - \|u_x\|^2 \left(1 - \frac{1}{2} \beta \|u(t)\| \right). \tag{2.20}$$

Next, assume that

$$\|u_0\| < 2\beta^{-1} \qquad \left(\text{i.e.,} \quad \int_0^1 u_0^2(x) \, dx < 4/\beta^2 \right).$$

Then either

(i) $\qquad\qquad\qquad \|u(t)\| < 2\beta^{-1}, \qquad \forall \, t > 0$

or

(ii) there exists an $\eta < \infty$ such that

$$\|u(\eta)\| = 2\beta^{-1}, \quad \text{with}$$
$$\|u(\eta)\| < 2\beta^{-1}, \quad \text{on} \quad [0, \eta).$$

Suppose (ii) holds. Then on $[0, \eta)$, $\quad 1 - \frac{1}{2}\beta\|u(t)\| > 0$ so (2.20) shows

$$\frac{d}{dt} \|u\|^2 < 0, \quad \text{for} \quad 0 \leq t < \eta. \tag{2.21}$$

Hence,

$$\|u(t)\|^2 \leq \|u(0)\|^2 = \|u_0\|^2 < 4\beta^{-2}, \qquad t \in [0, \eta).$$

Since $\|u(t)\|$ is assumed continuous in t, this means $\|u(\eta)\| \neq 2/\beta$, a contradiction. Hence, (ii) is false and (i) holds. (We are assuming the solutions we are dealing with are "classical", and so $u \in C^2$ in x, $u \in C^1$ in t.) Therefore, provided

$$\|u_0\| < 2/\beta,$$

it follows that

$$\|u(t)\| < 2/\beta, \qquad \forall\, t \geq 0.$$

Further, (2.21) now holds $\forall\, t \geq 0$, and hence

$$\|u(t)\|^2 \leq \|u_0\|^2, \qquad \forall\, t \geq 0.$$

We have shown that

$$1 - \frac{1}{2}\beta\|u(t)\| \geq 1 - \frac{1}{2}\beta\|u_0\| \ (> 0).$$

Now, use this in (2.20),

$$\frac{1}{2}\frac{d}{dt}\|u\|^2 \leq -\|u_x\|^2\left(1 - \frac{1}{2}\beta\|u(t)\|\right),$$
$$\leq -\|u_x\|^2\left(1 - \frac{1}{2}\beta\|u_0\|\right).$$

Next, from Poincaré's inequality, $\|u_x\|^2 \geq \pi^2\|u\|^2$, and since $(1 - \frac{1}{2}\beta\|u_0\|) > 0$, we find

$$\frac{1}{2}\frac{d}{dt}\|u\|^2 \leq -\pi^2\left(1 - \frac{1}{2}\beta\|u_0\|\right)\|u\|^2$$
$$= -A\|u\|^2,$$

where we have set $A = \pi^2(1 - \frac{1}{2}\beta\|u_0\|)$.

Using an integrating factor we may obtain from this inequality

$$\frac{d}{dt}(\|u(t)\|^2 e^{2At}) \leq 0,$$

which in turn integrates to

$$\|u(t)\|^2 \leq e^{-2At}\|u_0\|^2.$$

What we have shown is that if $\|u_0\| < 2/\beta$, then $\|u(t)\| \to 0$ at least exponentially fast. We refer to this as *nonlinear conditional stability*. Nonlinear since the nonlinear terms are handled in the analysis and conditional since the initial data has to be small enough; although it is important to note that we find a threshold for u_0 even though it is not necessarily the best (i.e., it is possible $u \to 0$ even if $\|u_0\| \geq 2\beta^{-1}$).

2.4 Weighted Energy and Boundedness for the Quadratically Forced Diffusion Equation

In the last section we saw that if $\|u_0\| < 2\beta^{-1}$, then a solution to (2.17) will decay to zero in L^2 norm, at least exponentially fast. It is natural to wonder what will happen if this restriction on the initial data is exceeded, especially as when the convective uu_x term is not present, a u^2 term can lead to global nonexistence of the solution in finite time by some norm of the solution blowing up, see e.g., Straughan (1982) and the references therein.

In this section we shall show by using an L^2 "energy" with a weight that finite time blow-up for a solution to (2.17) cannot occur, at least not in L^2 norm. The technique is simple but appealing as it uses the convective term to stabilize the problem. The material of this section and §§(2.5), (2.6) is very much related to the paper of Levine et al. (1989).

Let u be a solution to the boundary-initial value problem (2.17), which for clarity we rewrite here,

$$
\begin{aligned}
&\frac{\partial u}{\partial t} + u\frac{\partial u}{\partial x} = \frac{\partial^2 u}{\partial x^2} + \beta u^2, \qquad x \in (0,1),\ t > 0, \\
&u(0,t) = u(1,t) = 0, \qquad \forall\, t \geq 0, \\
&u(x,0) = u_0(x).
\end{aligned}
\tag{2.17}
$$

We shall restrict attention to positive solutions, $u(x,t) \geq 0$, which we might expect to be most destabilizing. Furthermore, although we consider any positive β, we are primarily interested in β large, when the conditional result becomes restrictive.

Let us introduce the notation $< \cdot >$ to mean the integral over (0,1), and let $\mu(x) = e^{-kx}$ for $k(> 0)$ to be defined, and then define

$$
F(t) = < \mu u^2 > = \int_0^1 \mu(x)u^2(x,t)\,dx.
\tag{2.22}
$$

Differentiate this with respect to t, then use (2.17), integrate by parts, and use the boundary conditions to find

$$
\begin{aligned}
F_t &= 2 < \mu u(-uu_x + u_{xx} + \beta u^2) > \\
&= 2\beta < \mu u^3 > + \frac{2}{3} < \mu_x u^3 > -2 < \mu u_x^2 > + < \mu_{xx} u^2 > \\
&= -2\left(\frac{k}{3} - \beta\right) < \mu u^3 > -2 < \mu u_x^2 > +k^2 < \mu u^2 > .
\end{aligned}
\tag{2.23}
$$

Now, by Hölder's inequality,

$$
< \mu u^2 > \leq < \mu u^3 >^{2/3} < \mu >^{1/3},
$$

or evaluating $< \mu >$ and rearranging,

$$- < \mu u^3 > \le - \frac{k^{1/2} < \mu u^2 >^{3/2}}{(1 - e^{-k})^{1/2}} . \qquad (2.24)$$

We also need the value for

$$\tilde{\lambda}_1 = \min_{H_0^1(0,1)} \frac{< \mu(x) u_x^2 >}{< \mu u^2 >} . \qquad (2.25)$$

The Euler equation for this problem is

$$u_{xx} - k u_x + \tilde{\lambda}_1 u = 0,$$

and substituting $u = e^{mx}$ we find

$$m = \frac{1}{2} k \pm i \sqrt{\tilde{\lambda}_1 - \frac{k^2}{4}} .$$

(The possibility that $m \in \mathbf{R}$ is quickly seen to lead to a zero solution.)
Hence,

$$u = A \exp\left(\frac{1}{2} kx\right) \cos \theta x \; + \; B \exp\left(\frac{1}{2} kx\right) \sin \theta x \;,$$

where $\theta = [\tilde{\lambda}_1 - k^2/4]^{1/2}$ give the extremals. The boundary conditions
require $A = 0$ and

$$\theta^2 = \pi^2, 4\pi^2, 9\pi^2, \ldots .$$

In consequence we find

$$\tilde{\lambda}_1 = \frac{k^2}{4} + \pi^2. \qquad (2.26)$$

We next choose $k = 3\beta + 1$, and use (2.24), (2.26) in (2.23) to obtain

$$F_t \le - \frac{2k^{1/2}}{3(1 - e^{-k})^{1/2}} F^{3/2} + \left(\frac{1}{2} k^2 - 2\pi^2\right) F.$$

(If $k^2 < 4\pi^2$, then from this we can deduce unconditional decay of F, i.e.,
for all u_0. However, we can do better than this, and this is shown in §2.5.)
To solve the above inequality divide by $F^{3/2}$, define $v = F^{-1/2}$ to obtain

$$\frac{dv}{dt} + Av \ge B,$$

where

$$A = \frac{k^2}{4} - \pi^2, \qquad B = \frac{k^{1/2}}{3(1 - e^{-k})^{1/2}} .$$

After using an integrating factor on the last inequality and then integrating, we solve for $F^{1/2}$ to see that

$$F^{1/2}(t) \leq \frac{1}{F^{-1/2}(0)e^{-At} + (B/A)(1 - e^{-At})} \ . \tag{2.27}$$

Without explicitly putting in A, B, it is easy to see that $F(t)$ is bounded for all t. As $t \to \infty$, the right hand side approaches

$$\frac{A}{B} = \left(\frac{k^2}{4} - \pi^2\right) \frac{3(1 - e^{-k})^{1/2}}{k^{1/2}} \sim O\left(\beta^{3/2}\right).$$

It follows that for large time (and large β),

$$F(t) \leq O\left(\beta^3\right).$$

Furthermore,

$$F(t) = <\mu u^2> \geq e^{-k} <u^2> \ .$$

Since $k = 3\beta + 1$, this yields the asymptotic behaviour

$$\|u(t)\|^2 \leq O\left(\beta^3 e^{3\beta}\right).$$

Note that $\|u\|$ can become very large, but is always bounded. In fact, for β above a threshold u tends to a positive steady state that has a steep boundary layer near $x = 1$. Numerical results showing the approach to steady state are shown in §2.8.

2.5 Weighted Energy and Unconditional Decay for the Quadratically Forced Diffusion Equation

The object of this section is to show that by modifying the argument of the last section we may obtain a value for β, such that for β less than this, all solutions to (2.17) decay (in a suitable norm) regardless of the size of the initial data.

Consider for now a positive, $u \geq 0$, solution to (2.17) and for a constant $\delta(> 0)$ to be specified choose

$$F(t) = <\mu u^{1+\delta}>, \tag{2.28}$$

where again $\mu = e^{-kx}$. By differentiating F, using the differential equation of (2.17), integrating by parts, and using the boundary conditions we find

$$F_t = (1 + \delta) <\mu u^\delta(-uu_x + u_{xx} + \beta u^2)>$$

$$= (1 + \delta)\beta <\mu u^{2+\delta}> + \left(\frac{1+\delta}{2+\delta}\right) <\mu_x u^{2+\delta}>$$

$$- \frac{4\delta}{(1+\delta)} <\mu w_x^2> + <\mu_{xx} u^{1+\delta}>,$$

where we have set $w = u^{(1+\delta)/2}$. Recalling the definition of μ, this leads to

$$F_t \leq (1 + \delta)\left(\beta - \frac{k}{2 + \delta}\right) < \mu u^{2+\delta} >$$
$$- \frac{4\delta}{(1+\delta)}\left(1 - \frac{k^2(1+\delta)}{4\delta}\tilde{\lambda}_1^{-1}\right) < \mu w_x^2 >, \tag{2.29}$$

where $\tilde{\lambda}_1^{-1} = \max_{H_0^1(0,1)} < \mu w^2 > / < \mu w_x^2 >$ and is given by (2.26), and we have used the fact that

$$< \mu_{xx} u^{1+\delta} > \leq k^2\left(\max_{H_0^1(0,1)} \frac{< \mu w^2 >}{< \mu w_x^2 >}\right) < \mu w_x^2 > .$$

If δ and k are now chosen such that

$$\frac{k^2(1+\delta)}{4\delta\tilde{\lambda}_1} < 1 \tag{2.30}$$

and $k = \beta(2 + \delta)$, then from (2.29) we use Poincaré's inequality to assert that $F(t) \to 0$, (at least) exponentially, as $t \to \infty$.

Now, recall from (2.26) that $\tilde{\lambda}_1 = \pi^2 + k^2/4$, then (2.30) and the choice for k require

$$\beta^2 < \frac{4\delta\pi^2}{(2+\delta)^2} = f(\delta), \quad \text{say.}$$

$f(\delta)$ achieves a maximum where $\delta = 2$ and so this is our optimum choice for δ in (2.28). With this choice, we deduce that provided

$$\beta^2 < \frac{1}{2}\pi^2, \tag{2.31}$$

solutions to (2.17) for arbitrarily large initial data decay to zero in the F measure of (2.28). Of course, from this we may also deduce $\|u\|_{L^2}$ decay with the aid of Hölder's inequality.

Finally in this section we observe that the argument with

$$F(t) = < \mu u^3 >$$

requires $u \geq 0$. To circumvent this we take

$$F(t) = < \mu u^{3+1/m} >,$$

m an odd integer; i.e., take a real root $u^{1/m}$ and then consider the non-negative quantity $u^{(3m+1)/m}$. Apply the argument of this section to this F and let $m \to \infty$ to see that (2.31) is sufficient for the decay of all solutions to (2.17), not just positive ones.

2.6 A Stronger Force in the Diffusion Equation

In §2.4 we saw that if the force in (2.17) is βu^2, then $\|u\|_{L^2}$ remains bounded for all t regardless of the size of β. It is natural to wonder if this result continues to hold for a larger power of u. The answer is no, as we now show.

Consider now the following boundary-initial value problem for the forced convection-diffusion equation,

$$
\begin{aligned}
\frac{\partial u}{\partial t} + u\frac{\partial u}{\partial x} &= \frac{\partial^2 u}{\partial x^2} + \beta u^{2+\delta}, \qquad x \in (0,1),\ t > 0, \\
u(0,t) &= u(1,t) = 0, \qquad \forall\, t \ge 0, \\
u(x,0) &= u_0(x),
\end{aligned}
\tag{2.32}
$$

where δ is *any positive* constant.

We shall show that it is possible to prescribe initial data for which no global solution exists. Let ϕ be the first eigenfunction in the membrane problem for (0,1), so here

$$\phi = \sin \pi x,$$

and consider for n to be determined

$$F(t) =< \phi^n u > . \tag{2.33}$$

By direct calculation and use of (2.32),

$$
\begin{aligned}
F_t &=< \phi^n u_t > \\
&=< \phi^n(-uu_x + u_{xx} + \beta u^{2+\delta}) > \\
&= -\pi^2 nF + n(n-1) < \phi^{n-2}\phi_x^2 u > +\beta < \phi^n u^{2+\delta} > +\frac{n}{2} < \phi^n \phi_x u^2 >,
\end{aligned}
$$

where the definition of ϕ has also been employed. From the last equation we may deduce that

$$F_t \ge -n\pi^2 F + \beta < \phi^n u^{2+\delta} > -\frac{1}{2}n\pi < \phi^{n-1}u^2 > . \tag{2.34}$$

Since from Hölder's inequality

$$< \phi^n u^{2+\delta} >^{2/(2+\delta)} \cdot < 1 >^{\delta/(2+\delta)} \ \ge \ < \phi^{2n/(2+\delta)}u^2 >,$$

the choice $n = 1 + 2/\delta$ gives

$$< \phi^{1+2/\delta}u^{2+\delta} > \ \ge \ < \phi^{2/\delta}u^2 >^{(2+\delta)/2},$$

and this in (2.34) allows us to see that

$$
\begin{aligned}
F_t \geq & -\left(\frac{\delta+2}{\delta}\right)\pi^2 F \\
& + \beta < \phi^{2/\delta}u^2 >^{(2+\delta)/2} -\frac{1}{2}\pi\left(\frac{\delta+2}{\delta}\right) < \phi^{2/\delta}u^2 > .
\end{aligned}
\tag{2.35}
$$

Further, by the Cauchy-Schwarz inequality

$$
< \phi^{2/\delta}u^2 > \; \geq \; \frac{< \phi^{(\delta+2)/\delta}u >^2}{< \phi^{2(\delta+1)/\delta} >} \; \geq \; < \phi^{(\delta+2)/\delta}u >^2 .
$$

Use of this on the first term on the right in (2.35) gives

$$
F_t \geq -\left(\frac{\delta+2}{\delta}\right)\pi^2 G^{\frac{1}{2}} + \beta G^{(2+\delta)/2} - \frac{1}{2}\pi\left(\frac{\delta+2}{\delta}\right)G,
\tag{2.36}
$$

where

$$
G(t) =< \phi^{2/\delta}u^2 > .
$$

Let us further assume that $u_0(x)$ is chosen such that

$$
\beta F(0)^{2+\delta} - \frac{1}{2}\pi\left(\frac{\delta+2}{\delta}\right)F^2(0) - \left(\frac{\delta+2}{\delta}\right)\pi^2 F(0) \; > \; 0.
\tag{2.37}
$$

Then by continuity there exists a t_1 such that for $0 < t < t_1$,

$$
R(F) = \beta F^{2+\delta}(t) - \frac{1}{2}\pi\left(\frac{\delta+2}{\delta}\right)F^2(t) - \left(\frac{\delta+2}{\delta}\right)\pi^2 F(t) \; > \; 0.
$$

For t in this range we check that $R'(F) > 0$, and so since $G^{1/2} \geq F$, $R(G^{1/2}) \geq R(F)$, for $0 \leq t < t_1$. Hence, in this t-interval we may replace G on the right of (2.36) by F to find

$$
F_t \geq R(F).
$$

Since F' is increasing on $(0, t_1)$ and $R(F)$ is also increasing on $(0, t_1)$, $R(F(t_1)) \neq 0$. Hence, t_1 must either be infinity or the limit of existence of the solution. Therefore, separating variables,

$$
\infty > \int_{F(0)}^{\infty} \frac{dF}{R(F)} \geq \int_{F(0)}^{F(t_1)} \frac{dF}{R(F)} \geq t_1 .
$$

This leads to a contradiction if t_1 were infinite, and we deduce the solution ceases to exist globally in a finite time.

The behaviour expected is blow-up of the solution in finite time, as indeed it is, see Levine et al. (1989). While the material of this section is not strictly in keeping with the rest of the book, we include it since it is important to realize that global existence need not follow. Convection problems with non-Boussinesq or phase change effects can lead to quadratic and higher nonlinearities, and this simple example indicates care has to be exercised.

2.7 A Polynomial Heat Source in Three Dimensions

So far we have concentrated on examples in one spatial dimension. We here introduce a different *energy* that allows us to produce a decay result for an arbitrary polynomial type source term, provided we suitably restrict the initial data. It also allows us to see treatment of a problem in three dimensions, for which the "embedding" inequalities are different from those in one dimension.

In this section we study the asymptotic behaviour of a solution u to the boundary-initial value problem,

$$
\begin{aligned}
\frac{\partial u}{\partial t} &= \Delta u + f(u), & \text{in} \quad & \Omega \times (0, \infty), \\
u &= 0, & \text{on} \quad & \Gamma \times (0, \infty), \\
u &= u_0(\mathbf{x}), & t = 0, \quad & \mathbf{x} \in \bar{\Omega},
\end{aligned}
\tag{2.38}
$$

where Ω is a bounded domain in $\mathbf{R}^3$ with (sufficiently smooth) boundary Γ.

The nonlinear function f satisfies the conditions $f(0) = 0$ and

$$
f(u) \leq \gamma u^k,
\tag{2.39}
$$

for $u \geq 0$, constant γ and k with $\gamma > 0$, $k > 1$. (For $k > 1$ it is known that unless the initial data is small enough the solution u can blow up in finite time, see e.g., Levine (1973), Straughan (1982).)

Again we employ an energy argument and the bound on the initial data is analogous to the bound required on the initial energy in §2.3, where only conditional asymptotic stability is found. The *energy* we employ is no longer the L^2-integral of u. To dominate the arbitrarily high power of u in $f(u)$, we find it necessary to use a higher power of u as energy: this was suggested by Bailey et al. (1984).

Henceforth, we suppose $u \geq 0$ and $k > 5/3$. As we are primarily interested in the decay behaviour for large k this causes no loss in generality.

Theorem 2.1.
Suppose $u(\geq 0)$ is a solution to (2.38) with f satisfying (2.39). If

$$
\|u_0^p\| < \left(\frac{(3k - 5)\pi^{2/3}2^{7/3}}{3^{3/2}\gamma(k - 1)^2} \right)^{3/4},
\tag{2.40}
$$

where
$$4p = 3(k - 1), \tag{2.41}$$

then $\|u^p\| \to 0$ at least exponentially, as $t \to \infty$, where $\| \cdot \|$ denotes the norm on $L^2(\Omega)$.

Proof.

Define E_p by
$$E_p(t) = \|u^p(t)\|^2 = \int_\Omega u^{2p} \, dx \, ,$$

for p a positive constant. Then
$$\frac{dE_p}{dt} = 2p \int_\Omega u^{2p-1} u_t \, dx$$
$$= 2p \int_\Omega u^{2p-1} \big[\Delta u + f(u) \big] \, dx \, . \tag{2.42}$$

Define I_1 and I_2 by
$$I_1 = -2p(2p - 1) \int_\Omega u^{2p-2} |\nabla u|^2 \, dx,$$

$$I_2 = 2\gamma p \int_\Omega u^{2p+k-1} \, dx.$$

Then note that since $u = 0$ on Γ,
$$2p \int_\Omega u^{2p-1} \Delta u \, dx = I_1, \tag{2.43}$$

and from (2.39),
$$2p \int_\Omega u^{2p-1} f(u) \, dx \leq I_2. \tag{2.44}$$

Use of (2.43) and (2.44) in (2.42) shows that
$$\frac{dE_p}{dt} \leq I_1 + I_2. \tag{2.45}$$

Next, I_1 may be rearranged as
$$I_1 = -\frac{2}{p}(2p - 1) \|\nabla u^p\|^2. \tag{2.46}$$

Furthermore, with the aid of Hölder's inequality,
$$I_2 \leq 2p\gamma \|u^{3p}\|^{2/3} \|u^{3(k-1)/4}\|^{4/3}. \tag{2.47}$$

Since $u = 0$ on Γ,

$$\|u^{3p}\|^{2/3} \leq \frac{2^{2/3}}{\pi^{2/3}3^{1/2}}\|\nabla u^p\|^2,$$

where the Sobolev embedding of $W_0^{1,2}(\Omega) \subset L^6(\Omega)$ has been employed (see e.g., Gilbarg & Trudinger (1977) or Appendix 1).

The last inequality is substituted into (2.47) and the resulting inequality is used together with (2.46) in (2.45); then if p is chosen as in (2.41) we find

$$\frac{dE_p}{dt} \leq -2\|\nabla u^p\|^2 \left(\frac{2p-1}{p} - \frac{2^{2/3}p\gamma}{\pi^{2/3}3^{1/2}} E_p^{2/3}\right). \qquad (2.48)$$

Since (2.40) and (2.41) hold, the coefficient of $-2\|\nabla u^p\|^2$ is positive, at least in an open neighbourhood of 0, say $[0,\eta)$. If $\eta \neq \infty$, and since (2.48) shows E_p decreases on $[0,\eta)$, a contradiction is obtained. Hence, (2.40) ensures

$$\frac{dE_p}{dt} \leq -2\left(\frac{2p-1}{p} - \frac{2^{2/3}p\gamma}{\pi^{2/3}3^{1/2}} E_p^{2/3}(0)\right)\|\nabla u^p\|^2.$$

Denote the coefficient of $-2\|\nabla u^p\|^2$ in the above by A, and since

$$\|\nabla u^p\|^2 \geq \xi_1 E_p,$$

(Poincaré inequality for Ω) for $\xi_1 > 0$, we find

$$\frac{dE_p}{dt} \leq -2A\xi_1 E_p,$$

and the theorem is proved.

The key to the theorem is the bound (2.40), which is a testable criterion since $u_0(x)$ is a measurable quantity. It should be observed that it depends on both $u_0(x)$ and Ω, and so it is a restriction on the size of both the initial data and the domain.

2.8 Sharp Conditional Stability

In §§2.3 and 2.7 we have seen that it is possible, for solutions to the non-linear parabolic equations considered there, to obtain conditional decay (stability) results. In their nonlinear stability analysis of the problem of Couette flow between concentric cylinders Joseph & Hung (1971) derived very sharp estimates on the Reynolds number, although they too found it necessary to restrict the size of the initial kinetic energy, thereby introducing the concept of conditional stability into the theory of nonlinear energy stability in fluid mechanics as a means of obtaining very sharp *quantitative* results. Since then, it has been found that very sharp nonlinear stability results may also be obtained in various thermal or salt convection problems of practical interest (some of these problems are discussed in chapters 4 to 7). However, some of this work, like Joseph & Hung (1971), obtains sharp conditions on the Rayleigh number at the expense of restricting the size of the amplitude of the initial perturbation. One of the open problems in these cases of energy stability theory is to find out what happens when the Rayleigh number is less than the *energy limit*, but in excess of the initial energy restriction. It would appear that answering the questions outlined above will require a clever choice of *generalized energy* (Lyapunov functional) or will require computation of three-dimensional flows for an incompressible fluid with highly nonlinear effects taking place.

We now present a simple but relevant example where the conditions on the Rayleigh number and initial energy *must* be used together to determine the subsequent solution behaviour. This example is taken from Galdi & Straughan (1987).

On page 458 of Drazin & Reid (1981) they use a nonlinear Burgers' equation to illustrate a bifurcation example. Here we study a modification of their problem, namely,

$$\frac{\partial u}{\partial t} + \alpha u \frac{\partial u}{\partial x} = R^{-1}\frac{\partial^2 u}{\partial x^2} + u + R\|u\|^\epsilon u, \qquad x \in (0,1), \ t > 0, \qquad (2.49)$$

where R is a positive parameter, ϵ is a positive constant, α is 1 or 0, and $\|\cdot\|$ again denotes the norm on $L^2(0,1)$. The solution u satisfies homogeneous boundary conditions

$$u(0,t) = u(1,t) = 0, \qquad \forall \, t \geq 0, \qquad (2.50)$$

and initial data

$$u(x,0) = u_0(x), \qquad x \in (0,1). \qquad (2.51)$$

We first establish

Theorem 2.2.
For $\alpha = 0$ or 1, if

$$\|u_0\|^\epsilon \quad < \quad C(R) \equiv \frac{(\pi^2 - R)}{R^2}, \qquad (2.52)$$

and $R < \pi^2$, then

$$\|u\| \to 0, \quad t \to \infty.$$

Proof.

Using (2.49), (2.50), the energy equation is

$$\frac{1}{2}\frac{d}{dt}\|u\|^2 = -R^{-1}\|u_x\|^2 + \|u\|^2 + R\|u\|^{2+\epsilon},$$

$$\leq -\|u\|^2\left(\frac{\pi^2}{R} - 1 - R\|u\|^\epsilon\right),$$

where in the last step Poincaré's inequality has been employed. If now $R < \pi^2$ and (2.52) holds, the theorem follows directly from this inequality by an argument similar to that of §2.3.

We now show that (2.52) is sharp in the sense that if it is violated there is non-uniqueness of the steady solution, while a catastrophic instability may result in the time-dependent problem; of course, it must be observed that this is the situation for which $u \equiv 0$ is judged to be *linearly* stable.

The case $\alpha = 0$, $R < \pi^2$.

Theorem 2.3.
Suppose $\alpha = 0$ and $R < \pi^2$.

(A) If $\|u_0\|^\epsilon = C(R)$, then there are at least two steady solutions to (2.49), (2.50).

(B) If $\|u_0\|^\epsilon > C(R)$, then there are an infinite number of steady solutions to (2.49), (2.50).

(C) If $\|u_0\|^\epsilon > [(2 + \epsilon)/2]C(R)$ and $u_0 = \sin \pi x$, then u ceases to exist in a finite time.

Proof.

Let $\bar{u} = A\sin n\pi x$. The steady form of equation (2.49) is

$$0 = R^{-1}\bar{u}_{xx} + B\bar{u},$$

where

$$B = 1 + R\|\bar{u}\|^\epsilon,$$

and so from these equations we must have

$$B = R^{-1}n^2\pi^2 \qquad \text{and} \qquad B = 1 + RA^\epsilon 2^{-\epsilon/2}.$$

Hence,

$$\frac{A^\epsilon}{2^{\epsilon/2}} \quad = \quad \|\bar{u}\|^\epsilon \quad = \quad \frac{n^2\pi^2 - R}{R^2}. \qquad (2.53)$$

Part (A) follows immediately from (2.53) with $n = 1$, since $\bar{u} = 0$ is a second solution.

Part (B) follows directly from (2.53), which defines $A(n)$.

To establish (C) we may use a concavity argument of Levine (1973): his abstract equation

$$Pu_t = -Au + \mathcal{F}(u),$$

covers (2.49) with $\alpha = 0$, provided

$$Au = -R^{-1}u_{xx} - u \quad \text{and} \quad \mathcal{F}(u) = Ru\|u\|^\epsilon.$$

The potential $\mathcal{G}(u)$ defined by Levine (1973) is here

$$\mathcal{G}(u) = \frac{R}{2+\epsilon}\|u\|^{2+\epsilon}.$$

The conditions of theorem 1 of Levine (1973), which establishes nonexistence, are satisfied provided

$$(u, Au) \geq 0 \tag{2.54}$$

and

$$\mathcal{G}(u_0) > \frac{1}{2}(u_0, Au_0), \tag{2.55}$$

the brackets denoting a suitable inner product; the inner product on $L^2(0,1)$ for the present example. Inequality (2.54) is equivalent to

$$R^{-1}\|u_x\|^2 \geq \|u\|^2;$$

this is certainly true if $R < \pi^2$.

Recalling that $u_0 = \sin \pi x$, inequality (2.55) is equivalent to

$$\|u_0\|^\epsilon \quad > \quad \left(\frac{2+\epsilon}{2}\right)C(R).$$

The theorem is thus proved.

We conclude this section with a brief examination of some numerical results for (2.49) when $\alpha = 1$ and associated equations.

Define the *Burgers' operator* Lu by

$$Lu = u_t + uu_x - R^{-1}u_{xx}. \tag{2.56}$$

We study three problems:

(I) $Lu = u$,

(II) $Lu = u + Ru\|u\|^\epsilon$,

(III) $Lu = u + Ru^2$,

where each of (I)–(III) are defined on $x \in (0,1)$, with $u \equiv 0$ on the boundaries $x = 0, 1$.

The numerical results were obtained with a finite element method using piecewise linear basis functions; further details of the numerical technique may be found in Straughan et al. (1987).

Problem I.
For $R < \pi^2$, the method of §2.4 shows $\|u\| \to 0$, $t \to \infty$. So, the interest is in the case $R > \pi^2$, for which u is *linearly* unstable, and the solution to the linearized problem grows exponentially.

When the uu_x term is present, we find that u does not tend to zero, nor does it grow indefinitely. Instead, it approaches a well defined "skew" steady state, with a boundary layer that is steeper the larger R.

The steady state and approaches to it are represented in Figures 2.1 to 2.6 below, where $R = 100$. The captions are self explanatory, but we note that for the initial data $u_0 = \sin 2\pi x$, $u_0 = \sin 3\pi x$, the solution can approach the positive steady state or a mirror image negative one, the actual one approached evidently being determined by numerical "noise". (In fact, with a very accurate scheme it is possible to see the solution tend to a steady state composed of multiples of the steady state shown, but repeated over smaller intervals, e.g., the intermediate "states" in figures 2.2, 2.5 show a "doubly steady state", and a "triply steady state", which eventually "flips over". For a smaller Δt, this "flip over" effect is much less easily seen and is a case where a very accurate numerical scheme looks at first sight to lead to too perfect solution behaviour.)

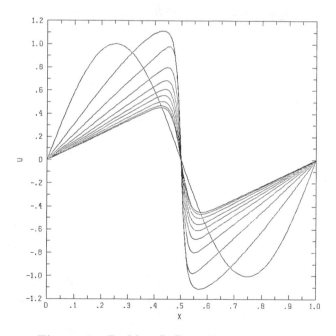

Figure 2.1 Problem I. $R = 100$, $u_0 = \sin 2\pi x$.
Plots at $t = .002, .2, .4, .6, .8, 1.0, 1.2, 1.5, 1.8, 2.0$, and the solution is always decaying and/or moving to the right.

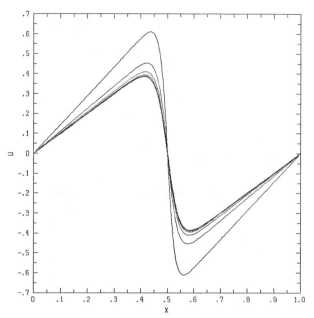

Figure 2.2 Problem I. $R = 100$, $u_0 = \sin 2\pi x$.
Plots at $t = 1, 2, ..., 10$. Solution still decaying.

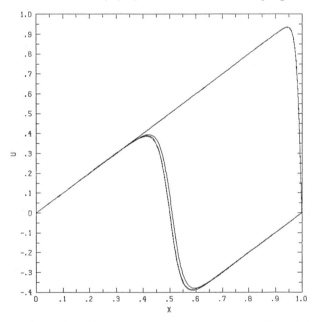

Figure 2.3 Problem I. $R = 100$, $u_0 = \sin 2\pi x$, $\Delta t = 0.05$.
Plots at $t = 5, 10, ..., 50$. Solution initially approaches a state
symmetric about $x = \frac{1}{2}$, before finally reaching
the true skew steady state.

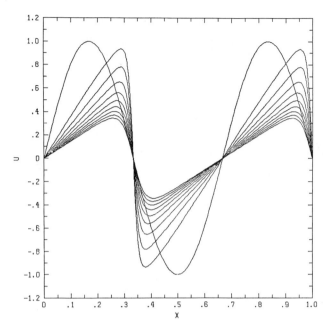

Figure 2.4 Problem I. $R = 100$, $u_0 = \sin 3\pi x$.
Plots at $t = .001, .2, .3, ..., 1.0$. Solution is decaying.

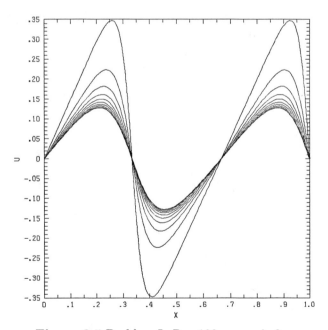

Figure 2.5 Problem I. $R = 100$, $u_0 = \sin 3\pi x$.
Plots at $t = 1, 2, ..., 10$. Solution approaching "triple" steady state.

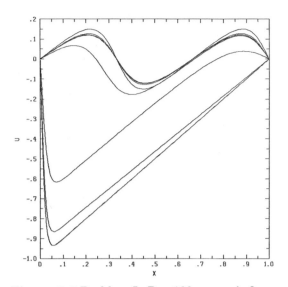

Figure 2.6 Problem I. $R = 100$, $u_0 = \sin 3\pi x$.
Plots at $t = 5, 10, ..., 50$. Solution finally goes to the (negative) steady state.

Problem II.

Here we show results for $R < \pi^2$. This is the situation where for the linearized problem $\|u\| \to 0$. For $\epsilon < 1$ we have been able to see a steady state approached for R not too large, with a steepening boundary layer for R, ϵ increasing. Typical plots are shown in Figures 2.7 to 2.9. For $\epsilon = 1$ we have

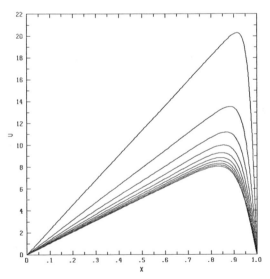

Figure 2.7 Problem II. $R = 2.5$, $\epsilon = 0.8$, $u_0 = 1000 \sin \pi x$.
Plots at $t = .2, .4, ..., 2$. Solution decaying to steady state with
$u_{\max} = 8.0601$.

not seen a steady state approached. Indeed, when $\epsilon = 1$, we have found it very difficult to proceed numerically beyond $t = 0.9$, and very rapid growth reminiscent of finite time blow-up is observed, cf. Straughan et al. (1987), Figures 8 and 9.

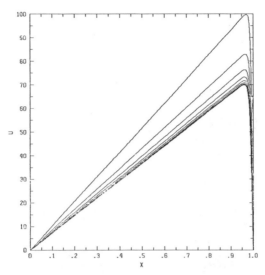

Figure 2.8 Problem II. $R = 2.1$, $\epsilon = 0.95$, $u_0 = 100 \sin \pi x$. Plots at $t = .2, .4, ..., 2$. Solution decaying to steady state with $x_{\max} = 0.962$, $u_{\max} = 70.1196$.

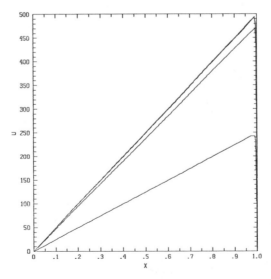

Figure 2.9 Problem II. $R = 2.3$, $\epsilon = 0.95$, $u_0 = 30 \sin \pi x$. Plots at $t = .2, .4, ..., 2$. Solution increasing to steady state with $x_{\max} = 0.994$, $u_{\max} = 494.9584$.

Problem III.

From the method of §2.4 we may show the solution to Problem III is always bounded. Figure 2.11 shows the final approach to steady state.

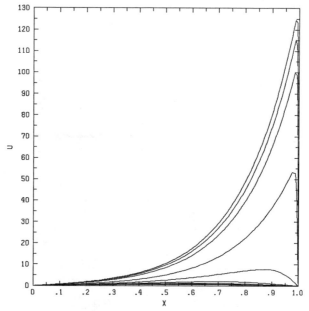

Figure 2.10 Problem III. $R = 5$, $u_0 = 0.5 \sin \pi x$.
Plots at $t = .1, .2, ..., .8, .83, .85$. Solution always growing.

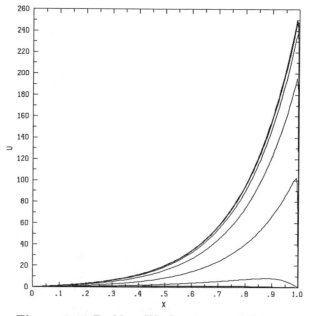

Figure 2.11 Problem III. $R = 5$, $u_0 = 0.5 \sin \pi x$.
Plots at $t = .2, .4, ..., 2$. Solution growing to a steady state.

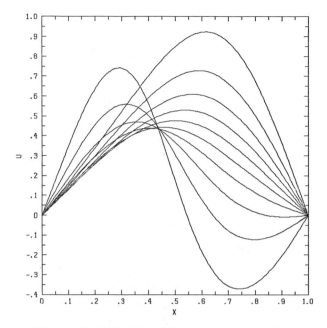

Figure 2.12 Problem III. $R = 5$, $u_0 = \sin 2\pi x$.
Plots at $t = .1, .2, ..., 1$. Showing solution in initial stages
of approach to steady state of Figure 2.11.

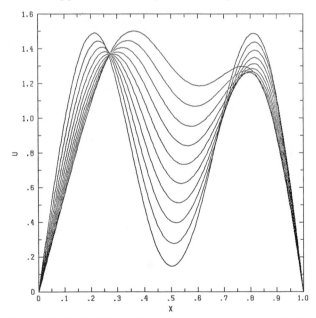

Figure 2.13 Problem III. $R = 5$, $u_0 = \sin \pi x + \sin 3\pi x$.
Plots at $t = .01, .02, ..., .1$. Showing solution in initial stages
of approach to steady state of Figure 2.11.

2.9 An Interaction Diffusion System

Up to this point we have considered examples for a single dependent variable satisfying one equation. As a prelude to fluid dynamical stability where the partial differential equations form a coupled system we here investigate how the techniques outlined in §§2.2 and 2.3 may be adapted to a system of interaction-diffusion equations of interest in many chemical and biological processes.

Consider two species of concentration C_1 and C_2 whose interaction and diffusion in a one-dimensional spatial region are governed by the equations,

$$\frac{\partial C_1}{\partial t} = D_1 \frac{\partial^2 C_1}{\partial x^2} + R_1(C_1, C_2),$$
$$\frac{\partial C_2}{\partial t} = D_2 \frac{\partial^2 C_2}{\partial x^2} + R_2(C_1, C_2). \tag{2.57}$$

In (2.57), D_i, $i = 1, 2$, are positive, constant diffusion coefficients, R_i are interaction terms, in general, nonlinear functions of C_1 and C_2.

Such a system may, for example, govern the evolutionary behaviour of an organic pollutant in water, in which case C_1 and C_2 might represent the pollutant and oxygen concentrations, respectively.

We suppose (2.57) are defined on $x \in \mathbf{R}$, $t > 0$. Differently from the equations studied earlier, (2.57) is a *system* linking C_1 and C_2 together. Our aim is to investigate using linear theory the stability of a *constant*, but in general *non-zero* solution to the time independent version of (2.57). This presupposes that the functions R_i are such that there exists a *constant* equilibrium solution $\bar{C}_1$, $\bar{C}_2$ to

$$R_1(\bar{C}_1, \bar{C}_2) = 0,$$
$$R_2(\bar{C}_1, \bar{C}_2) = 0. \tag{2.58}$$

To study the instability of the solution $(\bar{C}_1, \bar{C}_2)$ we let $c_1(x, t)$, $c_2(x, t)$ be perturbations so that in (2.57) we set

$$C_1 = \bar{C}_1 + c_1,$$
$$C_2 = \bar{C}_2 + c_2. \tag{2.59}$$

Next expand R_i in a Taylor series about $(\bar{C}_1, \bar{C}_2)$

$$R_i(C_1, C_2) = R_i(\bar{C}_1, \bar{C}_2) + \frac{\partial R_i}{\partial C_m}\Big|_{\bar{C}_j} c_m$$
$$+ \frac{1}{2} \frac{\partial^2 R_i}{\partial C_m \partial C_n}\Big|_{\bar{C}_j} c_m c_n + O(c^3), \tag{2.60}$$

where a repeated index denotes summation over 1 and 2. For linear stability, $|c_i| \ll 1$, so we neglect terms like $c_m c_n$ and higher order.

Define the (constant) matrix (a_{ij}) by

$$a_{im} = \left.\frac{\partial R_i}{\partial C_m}\right|_{\bar{C}_j},$$

then, since $R_i(\bar{C}_j) = 0$, (2.60) is, in the linear approximation

$$R_i(C_1, C_2) = a_{im}c_m. \tag{2.61}$$

Equations (2.57) for c_1, c_2 become

$$\frac{\partial c_1}{\partial t} = D_1 \frac{\partial^2 c_1}{\partial x^2} + a_{11}c_1 + a_{12}c_2,$$
$$\frac{\partial c_2}{\partial t} = D_2 \frac{\partial^2 c_2}{\partial x^2} + a_{21}c_1 + a_{22}c_2. \tag{2.62}$$

(We have already studied in §2.2 the case for $a_{12} = a_{21} = 0$, which would give linear instability always for $a_{11}, a_{22} > 0$.)

Now use a normal mode analysis, i.e., write

$$c_1 = \gamma_1 e^{(\sigma t + ikx)},$$
$$c_2 = \gamma_2 e^{(\sigma t + ikx)}, \tag{2.63}$$

with γ_i, $k \in \mathbf{R}$, and σ possibly complex. (Again we are looking at one term in the Fourier series for c_1, c_2, just as in the analysis for the diffusion equation.)

Putting (2.63) into (2.62) yields

$$\sigma\gamma_1 = -D_1 k^2 \gamma_1 + a_{11}\gamma_1 + a_{12}\gamma_2,$$
$$\sigma\gamma_2 = -D_2 k^2 \gamma_2 + a_{21}\gamma_1 + a_{22}\gamma_2, \tag{2.64}$$

which is a pair of simultaneous equations in γ_1, γ_2 that may be rewritten as

$$\begin{pmatrix} a_{11} - D_1 k^2 - \sigma & a_{12} \\ a_{21} & a_{22} - D_2 k^2 - \sigma \end{pmatrix} \begin{pmatrix} \gamma_1 \\ \gamma_2 \end{pmatrix} = \mathbf{0}.$$

For a non-zero solution (γ_1, γ_2) to this system, we require

$$\begin{vmatrix} a_{11} - D_1 k^2 - \sigma & a_{12} \\ a_{21} & a_{22} - D_2 k^2 - \sigma \end{vmatrix} = 0, \tag{2.65}$$

and, therefore, letting

$$\hat{a}_{11} = a_{11} - D_1 k^2,$$
$$\hat{a}_{22} = a_{22} - D_2 k^2,$$

(2.65) becomes

$$\sigma^2 - \sigma(\hat{a}_{11} + \hat{a}_{22}) + \hat{a}_{11}\hat{a}_{22} - a_{12}a_{21} = 0.$$

Hence,

$$\sigma = \frac{1}{2}(\hat{a}_{11} + \hat{a}_{22}) \pm \frac{1}{2}\left[(\hat{a}_{11} + \hat{a}_{22})^2 + 4(a_{12}a_{21} - \hat{a}_{11}\hat{a}_{22})\right]^{1/2}. \qquad (2.66)$$

Exchange of Stabilities. In general

$$\sigma = \sigma_1 + i\sigma_2, \qquad \sigma_1, \sigma_2 \in \mathbf{R}.$$

If $\sigma_2 \neq 0 \Rightarrow \sigma_1 < 0$ it then is said that the Principle of *Exchange of Stabilities* holds.

Of course, if $\sigma_2 = 0$ always, i.e., $\sigma \in \mathbf{R}$, then exchange of stabilities always holds. It is useful to know when $\sigma \in \mathbf{R}$ (if at all).

If $\sigma_2 \neq 0$ and exchange of stabilities does not hold, then any instability in that case is called an *overstable oscillation*.

If $\sigma \in \mathbf{R}$, overstability can be neglected and the analysis is usually much simpler.

Rewriting (2.66) as

$$\sigma = \frac{1}{2}(\hat{a}_{11} + \hat{a}_{22}) \pm \frac{1}{2}\left[(\hat{a}_{11} - \hat{a}_{22})^2 + 4a_{12}a_{21}\right]^{1/2}, \qquad (2.67)$$

an important case of exchange of stabilities is easily seen to be when $a_{12} = a_{21}$, i.e., when the system is symmetric.
(More generally,

$$\sigma \in \mathbf{R} \qquad \text{if} \qquad \text{sgn } a_{12} = \text{sgn } a_{21} .$$

This is equivalent, however, to a_{ij} symmetric since if $a_{12} = \alpha a_{21}$ ($\alpha > 0$) then we multiply the second of (2.62) by α and the resulting system has a_{ij} symmetric.)

If $a_{12}a_{21} < 0$, then there is a possibility that $\sigma_2 \neq 0$, i.e., there is a chance of overstability.

Suppose now a_{ij} symmetric. Then $\sigma \in \mathbf{R}$ and as $\sigma > 0$ implies instability we see, from (2.67), with the positive sign (most destabilizing),

$$\sigma = \frac{1}{2}(a_{11} - k^2 D_1 + a_{22} - k^2 D_2) + \frac{1}{2}\left[(\hat{a}_{11} - \hat{a}_{22})^2 + 4a_{12}a_{21}\right]^{1/2}. \quad (2.67')$$

If $a_{11}, a_{22} > 0$, there is always instability since as $k^2 \to 0$, σ becomes positive (c.f. §2.2, the diffusion equation for $x \in \mathbf{R}$).

It follows that a disturbance of infinite (spatial) wavelength is always destabilizing (i.e., in the limit as $k^2 \to 0$).

When there was only one equation, e.g.,

$$\frac{\partial c_1}{\partial t} = D_1 \frac{\partial^2 c_1}{\partial x^2} + a_{11}c_1, \qquad x \in \mathbf{R}, \, t > 0,$$

then $a_{11} < 0$ would always lead to stability of the solution $c_1 \equiv 0$. Is an analogous result true for the system (2.62)? No. It is sufficient to give one example where the $a_{12}a_{21}$ term can overcome the negativity of a_{11}, a_{22} and lead to an instability. Suppose next, therefore, that

$$D_1 = D_2 = D \,(> 0), \qquad a_{11}, a_{22} < 0,$$

where the conditions on the a's are both stabilizing in the single equation case, and

$$a_{12}a_{21} > a_{11}a_{22} \,(> 0). \tag{2.68}$$

Then, (2.67′) reduces to

$$\sigma = \frac{1}{2}(a_{11} + a_{22}) - k^2 D + \frac{1}{2}\left[a_{11}^2 + a_{22}^2 + 4a_{12}a_{21} - 2a_{11}a_{22}\right]^{1/2}. \tag{2.69}$$

Using (2.68) we may deduce from (2.69),

$$\begin{aligned}
\sigma &> \frac{1}{2}(a_{11} + a_{22}) - k^2 D + \frac{1}{2}|a_{11} + a_{22}| + \epsilon_1, \\
&\geq \epsilon_1 - k^2 D,
\end{aligned} \tag{2.70}$$

for a positive ϵ_1. As $k^2 \to 0$, σ becomes positive, so infinitely long wavelength disturbances are again unstable.

It should be observed that this is in complete contrast to the single equation case for which $a_{11} < 0$ or $a_{22} < 0$, i.e., $a_{12} = a_{21} = 0$, is always stable.

A deeper analysis of (2.67) is possible, but we do not pursue this here. (A concrete example where the analysis is taken further may be found in Segel & Jackson (1972).)

Conditional Nonlinear Stability in a Two-Constituent System. Consider now two species of concentrations C_1 and C_2 whose interaction and diffusion in a three-dimensional region are governed by the equations (c.f. (2.57)),

$$\begin{aligned}
\frac{\partial C_1}{\partial t} &= D_1 \Delta C_1 + R_1(C_1, C_2), \\
\frac{\partial C_2}{\partial t} &= D_2 \Delta C_2 + R_2(C_1, C_2).
\end{aligned} \tag{2.71}$$

In the above, D_i, $i = 1, 2$, are again positive diffusion coefficients and R_i are interaction terms.

We now take (2.71) to be defined on a bounded domain $\Omega \subset \mathbf{R}^3$ for each instant of time and on the boundary assume

$$\begin{aligned}
C_i &= g_i, & \text{on} \quad \Gamma_1 \times [0, \infty), \\
\frac{\partial C_i}{\partial n} &= h_i, & \text{on} \quad \Gamma_2 \times [0, \infty),
\end{aligned} \tag{2.72}$$

where $\Gamma_1 \cup \Gamma_2 = \Gamma$, $\Gamma_1 \neq \emptyset$, and g_i and h_i are prescribed. In addition, the initial values of C_1 and C_2 are given.

We wish to generalize the method of §2.3 to study the nonlinear stability of the steady solution to (2.71), i.e., the stability of $\bar{C}_i$ where

$$D_1 \Delta \bar{C}_1 + R_1(\bar{C}_j) = 0,$$
$$D_2 \Delta \bar{C}_2 + R_2(\bar{C}_j) = 0, \qquad (2.73)$$

with $\bar{C}_j$ satisfying the boundary conditions (2.72).

Our analysis is appropriate only to quadratic R_i. We write $C_i = \bar{C}_i + c_i$ where c_i is a perturbation to $\bar{C}_i$ and then

$$R_i(C_j) = R_i(\bar{C}_j) + a_{im}c_m + b_{imn}c_m c_n, \qquad (2.74)$$

where $a_{im} = \partial R_i / \partial C_m \big|_{\bar{C}_j}$, $2b_{imn} = \partial^2 R_i / \partial C_m \partial C_n \big|_{\bar{C}_j}$, b_{imn} being constants and a_{ij} functions of $\bar{C}_k$.

From (2.71), (2.73), (2.74), c_i satisfy the equations

$$\frac{\partial c_\alpha}{\partial t} = a_{\alpha m}c_m + b_{\alpha mn}c_m c_n + D_\alpha \Delta c_\alpha, \qquad (2.75)$$

where α signifies no summation although the summation convention applies to lower case Roman subscripts. Furthermore, c_i satisfy (2.72) with $g_i \equiv h_i \equiv 0$.

Let L be a length scale and introduce the non-dimensional variables

$$t^* = tM/L^2; \qquad M = D_1 + D_2; \qquad \mathbf{x} = L\mathbf{x}^*; \qquad \lambda_i = M/2D_i;$$
$$A_{\alpha j} = a_{\alpha j}L^2/D_\alpha; \qquad B_{\alpha mn} = b_{\alpha mn}L^2/D_\alpha.$$

The non-dimensional version of (2.75) is (omitting all stars)

$$2\lambda_\alpha \frac{\partial c_\alpha}{\partial t} = A_{\alpha m}c_m + B_{\alpha mn}c_m c_n + \Delta c_\alpha. \qquad (2.76)$$

We multiply (2.76) by c_α, integrate over Ω, and add the two equations to find

$$\frac{dE}{dt} = I - D + F(c), \qquad (2.77)$$

where

$$E(t) = \lambda_1 \|c_1\|^2 + \lambda_2 \|c_2\|^2, \qquad (2.78)$$

$$I = \int_\Omega A_{im}c_i c_m \, dx, \qquad D = \int_\Omega \frac{\partial c_i}{\partial x_q} \frac{\partial c_i}{\partial x_q} \, dx, \qquad (2.79)$$

in which the summation on i and m is over 1 and 2 whereas q sums from 1 to 3. Moreover,

$$F(c) = \int_\Omega B_{imn}c_i c_m c_n \, dx. \qquad (2.80)$$

We consider only the case A_{im} *symmetric.* For this case the idea of the proof leading to (2.67) shows the time factor σ is real and so the equations governing the *linear stability boundary* are

$$0 = A_{im}c_m + \Delta c_i. \tag{2.81}$$

These equations determine a set of relations between the a_{ij} and D_i.

To examine energy stability we return to (2.77) and derive

$$\frac{dE}{dt} \leq -D\left(1 - \max \frac{I}{D}\right) + F(c), \tag{2.82}$$

where the maximum is over the space of admissible solutions. We leave for the moment the determination of this maximum and suppose

$$k = -\max \frac{I}{D} + 1 > 0.$$

Denote by B the maximum of the constants B_{imn}. The nonlinear term is estimated using the Cauchy-Schwarz inequality

$$F(c) \leq 4\lambda B E^{1/2}\left[\left(\int_\Omega c_1^4 \, dx\right)^{1/2} + \left(\int_\Omega c_2^4 \, dx\right)^{1/2}\right], \tag{2.83}$$

where $\lambda = \max\{\lambda_1^{-1}, \lambda_2^{-1}\}$. Again, the Sobolev inequality $\|u^2\| \leq \mu\|\nabla u\|^2$ holds, although μ now depends on Ω (see Appendix 1). Using this in (2.83) leads to

$$F(c) \leq 4\lambda\mu B E^{1/2}D. \tag{2.84}$$

Therefore,

$$\frac{dE}{dt} \leq -kD + 4\lambda\mu B E^{1/2}D. \tag{2.85}$$

From this point the argument follows that after inequality (2.20) in §2.3, to show that if

$$E(0) < \frac{k^2}{16(\lambda\mu B)^2}, \tag{2.86}$$

then $E \to 0$, $t \to \infty$.

Thus, if $k > 0$ we have established a conditional stability result. It remains to investigate the condition $k > 0$.

We note that if the entries of A_{ij} are bounded, with $\Lambda = \max(I/D)$, the Euler-Lagrange equations for the maximum are

$$A_{im}c_m + \Lambda\Delta c_i = 0. \tag{2.87}$$

The limit of energy stability is $\Lambda = 1$; i.e., the boundary for $k > 0$, and for this case the Euler-Lagrange equations (2.87) of energy theory are exactly the same as those of the linear theory (2.81). Hence, the criteria for stability obtained by both methods are the same and so we conclude that when A_{ij} are symmetric and (2.86) holds the energy stability and linear instability boundaries coalesce and sub-critical instabilities do not occur. This point is taken up in further detail in chapter 4.

In the next chapter we turn to the Navier-Stokes equations and to the equations governing the behaviour of a layer of fluid heated from below.

3

The Navier-Stokes Equations, the Boussinesq Approximation, and the Standard Bénard Problem

3.1 Energy Stability for the Navier-Stokes Equations

For many flows it is sufficient to regard the fluid as incompressible. In this situation, with a constant temperature throughout, the velocity field, **v**, and pressure, p, are determined from the *Navier-Stokes* equations,

$$v_{i,t} + v_j v_{i,j} = -\frac{1}{\rho} p_{,i} + \nu \Delta v_i + f_i,$$

$$v_{i,i} = 0.$$

(3.1)

Here ρ is the (constant) density, ν (positive constant) is the kinematic viscosity, **f** is the external body force, and Δ is the Laplacian operator.

Notation. Standard indicial notation is used throughout together with the Einstein summation convention, and standard vector or tensor notation is used where appropriate. For example, in (3.1), $v_j v_{i,j} \equiv (\mathbf{v} \cdot \nabla)\mathbf{v}$ and $v_{i,i} \equiv \operatorname{div} \mathbf{v}$. A subscript t denotes partial differentiation with respect to time. When a superposed dot is used it denotes the material derivative, e.g.,

$$\dot{v}_i \equiv \frac{\partial v_i}{\partial t} + v_j v_{i,j}.$$

Henceforth throughout the book, Ω will denote a fixed, bounded region of 3-space with boundary, Γ, sufficiently smooth to allow applications of the divergence theorem. A major part of the book concerns motion in a plane layer, say $z \in (0, d)$. In this case, we usually refer to functions that have an (x, y) behaviour that is repetitive in the (x, y) direction, such as regular hexagons. The periodic cell defined by such a shape and its Cartesian product with $(0, d)$ will be denoted by V. The boundary of the period cell V will be denoted by ∂V.

The symbols $\| \cdot \|$ and $< \cdot >$ will denote, respectively, the L^2 norm on Ω and integration over Ω or the L^2 norm on V and integration over V : (for most of the book the domain is a period cell V) e.g.,

$$\int_V f^2 \, dV = \|f\|^2$$

and

$$\int_V fg \, dV = < fg > .$$

We shall study equations (3.1) on a bounded spatial region $\Omega (\subset \mathbf{R}^3)$ with $\mathbf{v}$ given on the boundary Γ, i.e.,

$$v_i(\mathbf{x}, t) = \hat{v}_i(\mathbf{x}, t), \qquad \mathbf{x} \in \Gamma. \tag{3.2}$$

Stability of the Zero Solution to (3.1) When $\mathbf{f}$ is Given by a Potential and $\hat{\mathbf{v}} \equiv \mathbf{0}$.
The physical situation here is one of a liquid in a container that is arbitrarily moved around and then at a preassigned time, say $t = 0$, the container is held fixed so that the only force acting is gravity. Then $\mathbf{f} = -\nabla\phi$, and $\mathbf{f}$ is given by a potential. Physically, we expect the existing disturbance to decay, no matter what its initial value.

Mathematically the problem is to investigate the behaviour of $\mathbf{v}$ that solves (3.1) with $\mathbf{f} = -\nabla\phi$, $\hat{\mathbf{v}} \equiv \mathbf{0}$ in (3.2) subject to the initial condition

$$v_i(\mathbf{x}, 0) = v_i^0(\mathbf{x}), \qquad \mathbf{x} \in \Omega. \tag{3.3}$$

Consider the (kinetic) energy $E(t)$ defined by

$$E(t) = \frac{1}{2}\|\mathbf{v}\|^2, \qquad \left(= \frac{1}{2}\int_\Omega v_i v_i \, dx \right). \tag{3.4}$$

We differentiate (3.4),

$$\frac{dE}{dt}(t) = < v_i v_{i,t} > .$$

Then from (3.1),

$$\frac{dE}{dt} = - < v_i v_j v_{i,j} > - < P_{,i} v_i > + \nu < v_i \Delta v_i >,$$

where we have put $P = p/\rho + \phi$. An integration by parts and use of the divergence theorem then yields

$$\frac{dE}{dt} = - \oint_\Gamma n_i P v_i \, dA + \int_\Omega P v_{i,i} \, dx$$

$$+ \nu \oint_\Gamma n_j v_i v_{i,j} \, dA - \nu \|\nabla \mathbf{v}\|^2$$

$$- \frac{1}{2} \oint_\Gamma n_j v_j v_i v_i \, dA + \frac{1}{2} \int_\Omega v_{j,j} v_i v_i \, dx .$$

Since $v_i \equiv 0$ on Γ, the boundary terms vanish. Further, from $(3.1)_2$, div $\mathbf{v} \equiv$ 0 and so we find

$$\frac{dE}{dt} = -\nu \|\nabla \mathbf{v}\|^2 . \tag{3.5}$$

From the Poincaré inequality (Appendix 1),

$$\lambda_1 \|\mathbf{v}\|^2 \leq \|\nabla \mathbf{v}\|^2 , \tag{3.6}$$

since here $\mathbf{v} \equiv \mathbf{0}$ on Γ, where $\lambda_1 > 0$. So use of (3.6) in (3.5) gives

$$\frac{dE}{dt} \leq -\nu \lambda_1 \|\mathbf{v}\|^2 = -2\nu \lambda_1 E(t). \tag{3.7}$$

Therefore,

$$E(t) \leq e^{-2\nu \lambda_1 t} E(0). \tag{3.8}$$

Hence, $\int_\Omega v_i v_i \, dx \to 0$ as $t \to \infty$.

We have shown that all disturbances to the zero solution, therefore, decay, as is physically expected. This is a classical result usually associated with Kampe de Feriet. Energy decay results of L^2 solutions to the Navier-Stokes equations on exterior domains pose a difficult question. For example, Poincaré's inequality no longer holds (in the form we have used it) and one can certainly not expect exponential decay in time. Answers to this mathematically technical problem are provided by Maremonti (1988), where further references are given. Maremonti's (1988) results are very sharp concerning the asymptotic decay rate.

We are usually interested in the stability of some particular *non-zero* flow. We indicate how one may develop an energy procedure for investigating the stability of a *non-zero* but *steady* solution to (3.1), (3.2).

Steady Solutions. Such a solution satisfies, with $P = p/\rho + \phi$,

$$\begin{aligned} V_j V_{i,j} &= -P_{,i} + \nu \Delta V_i, \\ V_{i,i} &= 0, \end{aligned} \tag{3.9}$$

in Ω, with

$$V_i(\mathbf{x}) = \hat{V}_i(\mathbf{x}), \tag{3.10}$$

on Γ, $\hat{V}_i$ being prescribed.

The idea is now to investigate the stability of $\mathbf{V}$. To do this we suppose $V_i^*(\mathbf{x}, t)$, $P^*(\mathbf{x}, t)$ is a solution to (3.1) that also satisfies the boundary data (3.10), and define

$$u_i(\mathbf{x}, t) = V_i^* - V_i , \tag{3.11}$$

where u_i is a perturbation to the steady velocity field $\mathbf{V}$. We write the pressure perturbation as p so that

$$p = P^* - P. \tag{3.12}$$

From (3.1) and (3.9) we then find u_i satisfies

$$u_{i,t} + u_j V_{i,j} + V_j^* u_{i,j} = -p_{,i} + \nu \Delta u_i,$$
$$u_{i,i} = 0, \tag{3.13}$$

in $\Omega \times (0, \infty)$, together with the homogeneous boundary data,

$$u_i = 0, \tag{3.14}$$

on $\Gamma \times [0, \infty)$.

The study of the stability of $\mathbf{V}$ is equivalent now to studying the stability of the zero solution ($\mathbf{u} \equiv \mathbf{0}$) to (3.13) subject to arbitrary initial disturbances,

$$u_i(\mathbf{x}, 0) = u_i^0(\mathbf{x}). \tag{3.15}$$

Nonlinear (Energy) Stability of $\mathbf{V}$. The idea is to determine a sufficient condition on $\mathbf{V}$ such that all disturbances $\mathbf{u}$ tend to zero as $t \to \infty$. No attempt is here made to optimize the analysis, since such refinements follow later.

Again consider the kinetic energy, but now of the perturbation $\mathbf{u}$,

$$E(t) = \frac{1}{2} \|\mathbf{u}\|^2. \tag{3.16}$$

Differentiate E and then substitute for $u_{i,t}$ from (3.13) to find

$$\frac{dE}{dt} = <u_i u_{i,t}>,$$
$$= - <u_i p_{,i}> + \nu <u_i \Delta u_i> - <V_j^* u_i u_{i,j}> - <u_j u_i V_{i,j}> \tag{3.17}$$

and denote the terms on the right of (3.17) by $I_1 - I_4$, respectively.

We consider $I_1 - I_4$ in turn. After integration by parts, use of the divergence theorem, and use of the solenoidal condition $(3.13)_2$ and the boundary condition (3.14), we derive the following equalities:

$$I_1 = - <u_i p_{,i}> = - \oint_\Gamma n_i u_i p \, dA + <p u_{i,i}> = 0.$$

$$I_2 = \nu <u_i \Delta u_i>$$
$$= -\nu <u_{i,j} u_{i,j}> + \oint_\Gamma \nu n_j u_i u_{i,j} \, dA$$
$$= -\nu \|\nabla \mathbf{u}\|^2.$$

$$I_3 = - <V_j^* u_i u_{i,j}> = -\frac{1}{2} <V_j^* (u_i u_i)_{,j}>$$
$$= -\frac{1}{2} \oint_\Gamma n_j V_j^* |\mathbf{u}|^2 \, dA + \frac{1}{2} <V_{j,j}^* u_i u_i>$$
$$= 0.$$

$$I_4 = - <V_{i,j} u_i u_j> = - <D_{ij} u_i u_j>,$$

where $D_{ij} = \frac{1}{2}(V_{i,j} + V_{j,i})$ is the symmetric part of the velocity gradient of the base solution. These relations in (3.17) yield

$$\frac{dE}{dt} = -\nu\|\nabla\mathbf{u}\|^2 - < D_{ij}u_ju_i > .\tag{3.18}$$

It is convenient now to non-dimensionalize this equation, i.e., to work with dimensionless variables. Let U be a "typical velocity" for the problem, and let L be a "typical lengthscale", e.g., L might be the maximum dimension of Ω, U may be $\max_\Omega |\mathbf{V}|$. Then since

$$\dim\nu \equiv [\nu] = \frac{L^2}{T} ,$$

we introduce,

$$\mathbf{x} = \hat{\mathbf{x}}L, \qquad t = \frac{L^2}{\nu}\hat{t}, \qquad \mathbf{u} = \hat{\mathbf{u}}U, \qquad \mathbf{V} = \hat{\mathbf{V}}U.\tag{3.19}$$

Note that the "hat" variables have no dimensions.

With these transformations the terms in (3.18) become

$$\frac{dE}{dt} = \frac{\nu U^2}{2L^2}L^3\frac{d}{d\hat{t}}\int_{\hat\Omega}\hat{u}_i\hat{u}_i\,d\hat{x} ,$$

$$\nu\int_\Omega u_{i,j}u_{i,j}\,dx = \frac{\nu U^2}{L^2}L^3\int_{\hat\Omega}\hat{u}_{i,j}\hat{u}_{i,j}\,d\hat{x} ,$$

$$\int_\Omega D_{ij}u_iu_j\,dx = \frac{U^3}{L}L^3\int_{\hat\Omega}\hat{D}_{ij}\hat{u}_i\hat{u}_j\,d\hat{x}.$$

Hence, the *dimensionless* form of (3.18) becomes (although for ease in writing we now *omit* all hats)

$$\frac{dE}{dt} = -\mathcal{D} - \frac{UL}{\nu}\int_\Omega D_{ij}u_iu_j\,dx ,\tag{3.20}$$

where

$$\mathcal{D} = \int_\Omega u_{i,j}u_{i,j}\,dx.\tag{3.21}$$

The dimensionless quantity UL/ν is denoted by Re, the *Reynolds number*. (After Osborne Reynolds who saw its relevance in stability theory — indeed, (3.20) is often called the Reynolds-Orr energy equation.) So, let

$$I = - < D_{ij}u_iu_j >,\tag{3.22}$$

then (3.20) is

$$\frac{dE}{dt} = -\mathcal{D} + RI,$$

$$\leq -\mathcal{D}R\Big(\frac{1}{R} - \max_{\mathcal{H}}\frac{I}{\mathcal{D}}\Big),\tag{3.23}$$

$\mathcal{H}$ being the space of admissible functions $\mathbf{u}$ over which the maximum is sought. Here, $\mathcal{H}$ is the space of functions

$$\mathcal{H} = \left\{ \mathbf{u} \middle| \mathbf{u} \in \left(H_0^1(\Omega) \right)^3, \, \nabla \cdot \mathbf{u} = 0 \right\}.$$

Denote

$$\frac{1}{R_E} = \max_{\mathcal{H}} \frac{I}{D}; \qquad (3.24)$$

then if $R < R_E$, $1/R - 1/R_E > 0$. Therefore, from (3.23)

$$\frac{dE}{dt} \leq -\left(\frac{R_E - R}{R_E} \right) \mathcal{D} \leq -2a\lambda_1 E, \qquad (3.25)$$

where we have also used Poincaré's inequality (3.6) and have set

$$a = \frac{R_E - R}{R_E} \, (> 0).$$

This may be integrated to yield

$$E(t) \leq e^{-2a\lambda_1 t} E(0). \qquad (3.26)$$

If $R < R_E$, then $E \to 0$ as $t \to \infty$ at least exponentially fast, and there is nonlinear stability of the steady solution $\mathbf{V}$.

The criterion of importance is (3.24), since if $R < R_E$, the base solution $\mathbf{V}$ is nonlinearly stable *for all initial disturbances* $\mathbf{u}_0$, regardless of how large they may be.

To calculate the maximum in (3.24), we use the calculus of variations. From a similar analysis to that leading to (2.15), we know that the maximizing solution satisfies

$$\delta I_2 - R_E \delta I_1 = 0, \qquad (3.27)$$

where now $I_2 = I_2(\nabla \mathbf{u})$, $I_1 = I_1(\mathbf{u})$. So,

$$\delta I_2 = \frac{d}{d\epsilon} I_2(\nabla \mathbf{u} + \epsilon \nabla \mathbf{h}) \bigg|_{\epsilon=0}$$

for all admissible $\mathbf{h}$ and

$$\delta I_1 = \frac{d}{d\epsilon} I_1(\mathbf{u} + \epsilon \mathbf{h}) \bigg|_{\epsilon=0}$$

for all admissible $\mathbf{h}$. Hence,

$$\delta I_2 = 2 < (u_{i,j} + \epsilon h_{i,j}) h_{i,j} > \big|_{\epsilon=0}$$
$$= 2 < u_{i,j} h_{i,j} >$$
$$= -2 < h_i \Delta u_i >,$$

since $\mathbf{h} = \mathbf{0}$ on Γ. Also,

$$\delta I_1 = 2 < D_{ij}(u_i + \epsilon h_i)h_j > \big|_{\epsilon=0}$$
$$= 2 < D_{ij}h_i u_j >.$$

However, since $\mathcal{H}$ is restricted to those functions that are divergence free, we must add into the maximum problem the constraint $u_{i,i} = 0$ by a Lagrange multiplier. This is done by adding a term

$$\int_\Omega \pi(\mathbf{x})u_{i,i}\, dx = 0$$

in the maximization: π depends on $\mathbf{x}$ since otherwise no contribution arises in the Euler-Lagrange equations from π. By adding in the above contribution, equation (3.27) reduces to

$$-2 < h_i(\Delta u_i + R_E D_{ij}u_j + \pi_{,i}) >= 0.$$

Since $\mathbf{h}$ is arbitrary, we need

$$\Delta u_i + R_E D_{ij}u_j = -\pi_{,i}\,,$$
$$u_{i,i} = 0\,, \tag{3.28}$$

in the region Ω together with the boundary conditions

$$u_i = 0, \quad \text{on } \Gamma. \tag{3.29}$$

As we want the *smallest* value of R_E, R_E is then determined as the lowest eigenvalue $R_E^{(1)}$ of (3.28), (3.29). (System (3.28), (3.29) is essentially a three-dimensional Sturm-Liouville problem. Since $\mathbf{V}(\mathbf{x})$ is known, (3.28), (3.29) is a *linear* problem for $\mathbf{u}^{(1)}$ and $R_E^{(1)}$.)

System (3.28), (3.29) for the energy eigenvalue R_E should be contrasted with the corresponding problem for the critical Reynolds number of linear theory. From (3.13), (3.14) the linearized equations for *linear instability* of the steady solution, $\mathbf{V}$, to (3.9), (3.10) are

$$\sigma u_i + u_j V_{i,j} + V_j u_{i,j} = -p_{,i} + \nu \Delta u_i,$$
$$u_{i,i} = 0,$$

in Ω, with

$$u_i = 0$$

on Γ. Here, σ is the (possibly complex) growth rate in the representation $u_i(\mathbf{x},t) = e^{\sigma t}u_i(\mathbf{x})$. If we use the non-dimensionalization (3.19), then this linearized system becomes

$$\sigma u_i + Re(u_j V_{i,j} + V_j u_{i,j}) = -p_{,i} + \Delta u_i,$$
$$u_{i,i} = 0, \tag{3.28L}$$

in Ω, with

$$u_i = 0 \qquad\qquad (3.29L)$$

on Γ, where Re is again the Reynolds number, $Re = UL/\nu$.

The logic is to determine the lowest eigenvalue R_E of (3.28), (3.29) and the corresponding lowest eigenvalue (value of Re), R_L, to (3.28L), (3.29L). If $R < R_E$ we are assured certain nonlinear stability, whereas if $R > R_L$ we know there is definitely instability. One of the main objectives of energy stability theory is to try to arrange that R_E is as close to R_L as possible. Of course, system (3.28L), (3.29L) is not symmetric (in a sense made precise in chapter 4), and is very different from the one of nonlinear energy stability theory, (3.28), (3.29). There are many problems in fluid mechanics, for example shear flow, where R_L and R_E are currently very different. (Professor S. Rionero, however, has informed me that he and Professor G. Mulone have recently constructed a *generalized energy* for the shear flow problem, which allows them to recapture many of the features of the linear theory by using a conditional nonlinear energy stability analysis.) In this book we describe several convection examples where by a suitable choice of an energy, it has been possible to arrange that R_E is very close to R_L and thus deliver results that are practically useful.

The determination of R_E for specific problems is considered later in the context of convection. A crude, but often useful, estimate may be obtained by directly deriving an upper bound for $I/\mathcal{D}$ from (3.23), instead of proceeding to (3.24), c.f. Serrin (1959a). From the spectral theorem of linear algebra we know that since $\mathbf{D}$ is a symmetric tensor,

$$I = - <D_{ij}u_i u_j> \leq \lambda_m \|\mathbf{u}\|^2 ,$$

where λ_m is the maximum of the three eigenvalues of $\mathbf{D}$, maximized over Ω. (In general, $D_{ij} = D_{ij}(\mathbf{x})$ so its eigenvalues too depend on $\mathbf{x}$, and λ_m denotes the maximum over Ω.)

Thus, from (3.23)

$$\frac{dE}{dt} \leq -\mathcal{D} + R\lambda_m \|\mathbf{u}\|^2$$
$$\leq -(\lambda_1 - R\lambda_m)\|\mathbf{u}\|^2, \qquad\qquad (3.30)$$

using also Poincaré's inequality (3.6).

Inequality (3.30) is

$$\frac{dE}{dt} \leq -2(\lambda_1 - R\lambda_m)E,$$

and this integrates to

$$E(t) \leq \exp\left\{-2(\lambda_1 - R\lambda_m)t\right\}E(0).$$

Thus,

$$R < \frac{\lambda_1}{\lambda_m}, \qquad (3.31)$$

represents a *sufficient* condition for nonlinear stability. Estimates for λ_1/λ_m are usually easy to determine, and may constitute a useful nonlinear stability estimate. In general, however, a criterion like (3.31) is much weaker than the variational result obtained from (3.28), (3.29).

3.2 The Balance of Energy and the Boussinesq Approximation

When we deal with motion of a fluid driven by buoyancy forces, such as those caused by heating the fluid, equations (3.1) are not sufficient and it is necessary to also add an equation for the temperature field, T. This is the equation for the *balance of energy*,

$$\rho T \dot{S} = \sigma_{ij} d_{ij} - q_{i,i}, \qquad (3.32)$$

where S, σ_{ij}, d_{ij}, and q_i are the entropy, stress tensor, symmetric part of the velocity gradient, and heat flux, and

$$\dot{S} = \frac{\partial S}{\partial t} + \mathbf{v}.\nabla S.$$

For a linear viscous fluid we take

$$\sigma_{ij} = -p\delta_{ij} + 2\mu d_{ij},$$

where μ $(= \rho\nu)$ is the dynamic viscosity. In general, the fluid is compressible, but for many convective motions the system may be considerably simplified by assuming the motion is isochoric, i.e., essentially incompressible flow, except in the body force term $\mathbf{f}$ in $(3.1)_1$. This approximation has caused a lot of concern in the literature and is known as the *Boussinesq approximation*.

In general, ρ is not constant and pressure variations have to be taken into account. Thus $S = S(\rho, p)$, and ρ, p, T are connected by an equation of state

$$f(\rho, p, T) = 0.$$

One may then show (cf. Batchelor (1967) pp. 164–171)

$$\rho T \dot{S} = \rho c_p \dot{T} - \alpha T \dot{p}, \qquad (3.33)$$

where c_p is the specific heat at constant pressure and

$$\alpha = -\frac{1}{\rho}\left(\frac{\partial \rho}{\partial T}\right)_p \qquad (3.34)$$

is the thermal expansion coefficient of the fluid. An order of magnitude argument like that of Batchelor (1967), pp. 164–171, may then be employed to show that provided

$$\frac{U^2}{c^2} \ll 1,$$ (3.35)

where U is a typical velocity in the problem and c is the local sound speed, the $\dot{p}$ term in (3.33) may be neglected as can the $\sigma_{ij}d_{ij}$ term in (3.32) by comparison with the other terms. Since for air at 15°C and one atmosphere pressure, $c = 340.6$ m sec^{-1} and, for water at the same temperature and pressure, $c = 1470$ m sec^{-1}, (3.35) is certainly satisfied in the convection problems we envisage here.

Thus, with a Fourier heat flux law,

$$\mathbf{q} = -k\nabla T,$$

(3.32) reduces to

$$\rho c_p \dot{T} = \nabla(k\nabla T).$$ (3.36)

We shall for now regard ρ, c_p, and k as constant. For many fluids this is very realistic.

The pressure in (3.1) is regarded as an unknown to be solved for in the problem, and the density is assumed constant except in the body force term $\mathbf{f}$, where

$$\rho\mathbf{f} = -g\mathbf{k}\rho(T),$$ (3.37)

assuming the only force acting is gravity, and where $\mathbf{k} = (0,0,1)$. The density in (3.37) is expanded in a Taylor series, and at this stage we consider only the first term, so

$$\rho = \rho_0\left[1 - \alpha(T - T_R)\right],$$ (3.38)

where ρ_0 is the density at temperature T_R, and α is the coefficient of thermal expansion given by (3.34).

Thus, taking account of (3.36)-(3.38), and (3.1), the equations for a linearly viscous, heat conducting, incompressible fluid (utilizing the Boussinesq approximation) are

$$v_{i,t} + v_j v_{i,j} = -\frac{1}{\rho_0}p_{,i} + \nu\Delta v_i - k_i g\left(1 - \alpha[T - T_R]\right),$$

$$v_{i,i} = 0,$$ (3.39)

$$T_{,t} + v_i T_{,i} = \kappa\Delta T,$$

where $\kappa = k/\rho_0 c_p$ is the thermal diffusivity.

For water between 0°C and 100°C (see e.g., Batchelor (1967) pp. 596, 597) c_p varies from 4.27 joule gm^{-1} deg C^{-1} to 4.216 joule gm^{-1} deg C^{-1},

κ varies from $1.33 \times 10^{-3}\,\mathrm{cm}^2\,\mathrm{sec}^{-1}$ to $1.66 \times 10^{-3}\,\mathrm{cm}^2\,\mathrm{sec}^{-1}$, so it is certainly reasonable to treat them as constant in many convection studies. The kinematic viscosity ν varies from $1.787 \times 10^{-2}\,\mathrm{cm}^2\,\mathrm{sec}^{-1}$ to $0.295 \times 10^{-2}\,\mathrm{cm}^2\,\mathrm{sec}^{-1}$ so its variation is greater. We shall look at thermal convection with ν constant, but there are practical situations where a strongly varying viscosity with temperature is necessary.

More elaborate treatments of the Boussinesq approximation are available and some of these are discussed in Joseph (1976b); see also Hills & Roberts (1990). However, they all basically arrive at a system like (3.39). Without a Boussinesq approximation to yield a simplified system like (3.39) with a solenoidal velocity field, one is left with treating convection in a compressible fluid, and this is a far more complicated issue.

An interesting new approach to the Boussinesq approximation is given in Hills & Roberts (1990) who adopt the principles of modern continuum thermodynamics, a philosophy to which we subscribe. A brief review of their novel approach, which also includes effects of compressibility in a precise manner, is now given.

Hills & Roberts (1990) begin with the full equations of motion for a compressible fluid in the form

$$\dot{\rho} = -\rho v_{i,i}\,,$$
$$\rho \dot{v}_i = \rho f_i + \sigma_{ji,j}\,,$$
$$\rho \dot{U} = \sigma_{ji} d_{ij} - q_{i,i} + \rho q, \qquad (3.40)$$

where q is the heat supply per unit mass per unit time and U is the specific internal energy. Their entropy production inequality is

$$\rho(T\dot{S} - \dot{U}) + \sigma_{ji} d_{ij} - \frac{q_i T_{,i}}{T} \geq 0. \qquad (3.41)$$

Their interest is in materials whose density can be changed by variations in the temperature, T, but not in the thermodynamic pressure P, and so they formulate the constitutive theory in terms of P and T. They argue that the natural thermodynamic potential is the Gibbs energy

$$G = U - ST + \frac{P}{\rho}.$$

The entropy inequality (3.41) may thus be rewritten

$$-\rho\big(\dot{G} + S\dot{T}\big) + \dot{P} + (\sigma_{ji} + P\delta_{ji})d_{ij} - \frac{q_i T_{,i}}{T} \geq 0. \qquad (3.42)$$

The constitutive theory of Hills & Roberts (1990) chooses

$$\begin{aligned} G = G(T,P), \qquad S = S(T,P), \qquad \rho = \rho(T), \\ \sigma_{ij} = -p\delta_{ij} + \lambda d_{mm}\delta_{ij} + 2\mu d_{ij}, \qquad q_i = -\kappa T_{,i}, \end{aligned} \qquad (3.43)$$

where p is the mechanical pressure, and the coefficients λ, μ, κ depend on P and T.

For the form of ρ in (3.43), the continuity equation (3.40)$_1$ becomes

$$\alpha \dot{T} = v_{i,i}, \tag{3.44}$$

where $\alpha (= -\rho^{-1} d\rho/dT)$ is the thermal expansion coefficient of the fluid. Equation (3.44) is regarded as a constraint and then included in (3.42) via a Lagrange multiplier, Γ. By using the arbitrariness of the body force and heat supply they then deduce from (3.42) that

$$S = -\left(\frac{\partial G}{\partial T} + \frac{\Gamma \alpha}{\rho}\right), \qquad \frac{\partial G}{\partial P} = \frac{1}{\rho},$$
$$p = P + \Gamma, \qquad \Gamma = \Gamma(T, P),$$
$$\lambda + \frac{2}{3}\mu \geq 0, \qquad \mu \geq 0, \qquad \kappa \geq 0.$$

They then work with another Gibbs energy, $\hat{G} = G + \Gamma/\rho$, which allows them to replace G, P, Γ by $\hat{G}, p$, for which

$$\hat{G} = \hat{G}(T, p) = G_0(T) + \frac{p}{\rho}, \qquad \frac{\partial \hat{G}}{\partial T} = -S, \qquad \frac{\partial \hat{G}}{\partial p} = \frac{1}{\rho},$$

where now the material parameters depend on p and T. The governing equations for what Hills & Roberts (1990) term a *generalized* incompressible linear viscous fluid become

$$\alpha \dot{T} = v_{i,i},$$
$$\rho \dot{v}_i = \rho f_i - p_{,i} + (\lambda v_{m,m} \delta_{ij} + 2\mu d_{ij})_{,j}, \tag{3.45}$$
$$\rho c_p \dot{T} - \alpha T \dot{p} = \rho q + (\kappa T_{,i})_{,i} + \lambda (d_{ii})^2 + 2\mu d_{ij} d_{ij},$$

where $c_p = T(\partial S/\partial T)_p$ is again the specific heat at constant pressure.

The boundary conditions they adopt are continuity of temperature and the normal component of the heat flux. At a stationary, no-slip boundary they have $v_i = 0$, while at a free surface, $\mathcal{S}$, at an ambient pressure π the stress vector is continuous, which results in

$$-\pi = -p + \lambda v_{i,i} + 2\mu n_i n_j v_{i,j}, \qquad \epsilon_{ijk} n_j \omega_k = 0, \qquad \text{on } \mathcal{S}, \tag{3.46}$$

n_i being the unit normal to $\mathcal{S}$ with ω_i the vorticity vector. In addition to (3.46) they have a kinematic condition

$$n_i v_i = V, \qquad \text{on } \mathcal{S}, \tag{3.47}$$

V being the velocity of $\mathcal{S}$ along its normal n_i.

This sytem must be further reduced to arrive at the usual equations, and so they consider a horizontal layer of fluid contained between a fixed lower boundary $z = L$ and a free top surface $z = f(x_1, x_2, t)$. The condition (3.47) is then

$$v_3 = \dot{f}, \qquad \text{on} \qquad z = f. \tag{3.48}$$

They assume gravity acts in the z-direction so $f_i = g\delta_{i3}$ and expand about a reference state (T_r, p_r) to obtain the expansion for the density

$$\rho(T) = \rho_r\left[1 - \alpha_r(T - T_r) + \cdots\right],$$

where $\rho_r = \rho(T_r)$, etc.

The key philosophy of Hills & Roberts (1990) (cf. Roberts (1967) pp. 196–197) is that *typical accelerations promoted in the fluid by variations in the density are always much less than the acceleration of gravity*. The resulting equations from the Boussinesq approximation, the so-called Oberbeck-Boussinesq equations, arise by taking the simultaneous limits, $g \to \infty$, $\alpha_r \to 0$, with the restriction that $g\alpha_r$ remains finite.

Hills & Roberts (1990) expand the pressure, velocity, and temperature fields in g^{-1}, viz.

$$p = p^0 g + p^1 + \frac{p^2}{g} + \cdots,$$

$$v_i = v_i^1 + \frac{v_i^2}{g} + \cdots,$$

$$T - T_r = T^1 - T_r + \frac{T^2 - T_r}{g} + \cdots,$$

together with an expansion of the free surface

$$f = \frac{f^2}{g} + \cdots.$$

Hills & Roberts (1990) derive the Oberbeck-Boussinesq equations at the $O(1)$ level with the limit

$$\epsilon = \frac{g\alpha_r L}{(c_p)_r} \to 0.$$

In fact, under a suitable non-dimensionalization, with

$$Ra = \frac{g\alpha_r \beta L^4}{\nu_r k_r}$$

being the Rayleigh number, and $Pr = \nu_r/k_r$ the Prandtl number, their $O(1)$ equations are

$$v_{i,i}^1 = 0,$$

$$\dot{v}_i^1 = -RaT^1\delta_{i3} - p_{,i}^1 + \Delta v_i^1, \tag{3.49}$$

$$Pr\left(\dot{T}^1 - \epsilon[T_r + T^1]v_3^1\right) = q + \Delta T + 2\frac{Pr}{Ra}\epsilon d_{ij}^1 d_{ij}^1.$$

Equations (3.39) are found when $\epsilon \to 0$ with $q \equiv 0$ in (3.49). It is important to observe that (3.49) do contain first order effects of compressibility via the ϵ terms.

3.3 Energy Stability and the Bénard Problem

In the majority of situations if a layer of fluid is heated from below the fluid in the lower part of the layer expands as it becomes hotter, and when the temperature gradient or layer depth is sufficiently large to overcome the effect of gravity the fluid rises and a pattern of cellular motion may be seen. This is called Bénard convection, after Bénard (1900); his experiments are now thought to have been driven by surface tension, see chapter 8. It is of historical interest to point out that the tesselated structure was previously observed by Thomson (1882). To describe this phenomenon mathematically we begin with equations (3.39).

We suppose the fluid is contained in the infinite layer $z \in (0, d)$ and the temperatures of the planes $z = 0, d$ are kept fixed,

$$T = T_0, \quad z = 0; \quad T = T_1, \quad z = d; \tag{3.50}$$

with $T_0 > T_1$. Then, the conduction solution (which is motionless) to (3.39), (3.50) is

$$\bar{\mathbf{v}} \equiv \mathbf{0}, \quad \bar{T} = -\beta z + T_0, \tag{3.51}$$

where

$$\beta = \frac{T_0 - T_1}{d},$$

and the pressure is determined from

$$\frac{d\bar{p}}{dz} = -\rho_0 g \big(1 - \alpha\{-\beta z + T_0 - T_R\}\big),$$

i.e.,

$$\bar{p} = -\rho_0 g \big[1 + \alpha(T_R - T_0)\big]z - \frac{\rho_0 g \alpha \beta}{2} z^2, \tag{3.52}$$

selecting the pressure scale to vanish at $z = 0$.

We wish to study the stability of the conduction solution (3.51), (3.52). To this end we introduce perturbations $\mathbf{u}, \theta, \pi$ to $\bar{\mathbf{v}}, \bar{T}$, and $\bar{p}$, respectively, i.e.,

$$\mathbf{v} = \bar{\mathbf{v}} + \mathbf{u}, \quad T = \bar{T} + \theta, \quad p = \bar{p} + \pi.$$

The perturbations are *not* assumed to be small. From (3.39) the perturbations are found to satisfy

$$u_{i,t} + u_j u_{i,j} = -\frac{1}{\rho_0} \pi_{,i} + \nu \Delta u_i + k_i g \alpha \theta,$$

$$u_{i,i} = 0, \tag{3.53}$$

$$\theta_{,t} + u_i \theta_{,i} = \beta w + \kappa \Delta \theta,$$

where $w = u_3$; in general, we write $\mathbf{u} = (u, v, w)$.

It is convenient to non-dimensionalize (3.53) according to the scales:

$$t = t^* \frac{d^2}{\nu}, \qquad \mathbf{x} = \mathbf{x}^* d, \qquad \theta = T^\sharp \theta^*, \qquad \mathbf{u} = \mathbf{u}^* U,$$

$$U = \frac{\nu}{d}, \qquad Pr = \frac{\nu}{\kappa}, \qquad T^\sharp = U\sqrt{\frac{\beta\nu}{\kappa\alpha g}},$$

$$R = \sqrt{\frac{\alpha g \beta d^4}{\kappa\nu}}, \qquad p^* P = \pi, \qquad P = \frac{\rho_0 \nu^2}{d^2}.$$

Here Pr is the Prandtl number and $Ra = R^2$ is the Rayleigh number. With this scaling the *non-dimensional* form of (3.53) becomes (we, as usual, omit all stars even though the *dimensionless* form is understood)

$$\begin{aligned} u_{i,t} + u_j u_{i,j} &= -p_{,i} + \Delta u_i + k_i R\theta, \\ u_{i,i} &= 0, \\ Pr(\theta_{,t} + u_i \theta_{,i}) &= Rw + \Delta\theta. \end{aligned} \qquad (3.54)$$

The boundary conditions on the planes $z = 0, 1$ are that there is no slip in velocity, and from (3.50) the temperatures remain constant, so

$$u_i = 0, \quad \theta = 0; \qquad z = 0, 1. \qquad (3.55)$$

In addition, we assume that $\mathbf{u}, \theta, p$ have an (x, y)-dependence consistent with one that has a repetitive shape that tiles the plane, such as two-dimensional rolls or hexagons. The hexagon solution was originally given by Christopherson (1940), namely,

$$u(x, y) = \cos\frac{1}{2}a(\sqrt{3}\,x + y) + \cos\frac{1}{2}a(\sqrt{3}\,x - y) + \cos ay. \qquad (3.56)$$

In particular, the (x, y)-dependence is consistent with a wavenumber, a, for which with

$$\Delta^* = \frac{\partial^2}{\partial x^2} + \frac{\partial^2}{\partial y^2},$$

u satisfies the relation

$$\Delta^* u = -a^2 u.$$

Whatever shape the cell has in the (x, y)-plane, its Cartesian product with $(0, 1)$ is the period cell V.

Before discussing nonlinear energy stability of a solution to (3.54) we briefly digress into linearized instability theory. The governing equations are obtained from (3.54) by omitting the nonlinear terms $u_j u_{i,j}$ and $Pr u_i \theta_{,i}$. The resulting linearized equations possess solutions of type

$$\mathbf{u}(\mathbf{x}, t) = \mathbf{u}(\mathbf{x})e^{\sigma t}, \qquad \theta(\mathbf{x}, t) = \theta(\mathbf{x})e^{\sigma t}, \qquad p(\mathbf{x}, t) = p(\mathbf{x})e^{\sigma t},$$

so that $\mathbf{u}(\mathbf{x}), \theta(\mathbf{x}), p(\mathbf{x})$ satisfy

$$\sigma u_i = -p_{,i} + \Delta u_i + k_i R\theta,$$
$$u_{i,i} = 0, \qquad\qquad (3.57)$$
$$\sigma Pr\theta = Rw + \Delta\theta.$$

The boundary conditions are still (3.55) and "periodicity" in the (x, y)-directions.

We now multiply $(3.57)_1$ by u_i^* (the complex conjugate of u_i), $(3.57)_3$ by θ^* and integrate over V, to find, integrating by parts and using the boundary conditions,

$$\sigma(< u_i u_i^* > + Pr < \theta\theta^* >) = R(< w\theta^* > + < w^*\theta >) \\ - (< \theta_{,i}\theta_{,i}^* > + < u_{i,j}u_{i,j}^* >). \qquad (3.58)$$

The right hand side of (3.58) is real and so if we let $\sigma = \sigma_r + i\sigma_1$, then taking the imaginary part of (3.58) we find

$$\sigma_1(\|\mathbf{u}\|^2 + Pr\|\theta\|^2) = 0,$$

the norm temporarily denoting the norm on the complex Hilbert space $L^2(V)$. Hence,

$$\sigma_1 = 0.$$

Therefore, in Bénard convection the growth rate σ is real. Thus, to find the instability boundary, the lowest value of R^2 in (3.57) for which $\sigma > 0$, we solve (3.57) for the smallest eigenvalue $R^2_{(1)}$ with $\sigma = 0$.

The linearized equations of Bénard convection satisfy the so-called *Principle of exchange of stabilities*. This is said to hold if in a system the growth rate σ is such that $\sigma \in \mathbf{R}$ or $\sigma_1 \neq 0 \Rightarrow \sigma_r < 0$. When this principle holds, convection sets in as *stationary convection*. If on the other hand, at the onset of instability, $\sigma = i\sigma_1$ with $\sigma_1 \neq 0$, the convection mechanism is referred to as *oscillatory convection*. It must be emphasized that the linearized theory only yields a boundary for *instability*, i.e., whenever $R > R_L$ the solution grows in time and is unstable (for (3.52) R_L is the lowest eigenvalue). In particular, the linearized equations do not yield any information on *nonlinear* stability: it is, in general, possible for the solution to become unstable at a value of R lower than R_L, and in this case subcritical instability (bifurcation) is said to occur. Details of how R_L^2 is found are given in Appendix 2, where numerical solutions of eigenvalue problems relevant to convection are described.

For the *standard Bénard problem*, (3.54), we shall now show by energy stability theory that sub-critical instability is not possible.

At this point we consider the simplest, natural "energy", formed by adding the kinetic and thermal energies of the perturbations, and so define

$$E(t) = \frac{1}{2}\|\mathbf{u}\|^2 + \frac{1}{2}Pr\|\theta\|^2. \qquad (3.59)$$

We differentiate E, substitute for $u_{i,t}, \theta_{,t}$ from (3.54), and use the boundary conditions to find

$$\frac{dE}{dt} = 2R < w\theta > - [D(\mathbf{u}) + D(\theta)], \qquad (3.60)$$

where $D(\cdot)$ denotes the Dirichlet integral, e.g.,

$$D(f) = \|\nabla f\|^2 = \int_V |\nabla f|^2 \, dV.$$

It is convenient to put

$$I = 2 < w\theta >, \qquad \mathcal{D} = D(\mathbf{u}) + D(\theta).$$

Then, from (3.60) we see that

$$\frac{dE}{dt} = RI - \mathcal{D} \leq -\mathcal{D}R\left(\frac{1}{R} - \frac{1}{R_E}\right), \qquad (3.61)$$

where R_E is defined by

$$\frac{1}{R_E} = \max_{\mathcal{H}} \frac{I}{\mathcal{D}}, \qquad (3.62)$$

with $\mathcal{H}$ being the space of admissible solutions.

If now

$$R < R_E, \qquad (3.63)$$

then

$$R\left(\frac{1}{R} - \frac{1}{R_E}\right) = \frac{R_E - R}{R_E} > 0;$$

and if we let

$$a = \frac{R_E - R}{R_E},$$

then from (3.61),

$$\frac{dE}{dt} \leq -a\mathcal{D} \leq -2a\lambda_1 E, \qquad (3.64)$$

where we have also used Poincaré's inequality. Inequality (3.64) easily integrates to obtain

$$E(t) \leq e^{-2a\lambda_1 t} E(0),$$

from which we see that $E \to 0$ at least exponentially fast as $t \to \infty$. This demonstrates that provided (3.63) is satisfied, the conduction solution (3.51) is nonlinearly stable for all initial disturbances.

The quantity of importance, therefore, is R_E as defined by (3.62). The Euler-Lagrange equations for this maximum are found by a similar procedure to that leading to (3.28), and we find

$$\Delta u_i + k_i R_E \theta = p_{,i},$$
$$u_{i,i} = 0, \qquad (3.65)$$
$$\Delta \theta + R_E w = 0,$$

together with the same boundary conditions as (3.55) and the "periodicity" conditions. It is immediately evident that R_E satisfies the same eigenvalue problem as (3.57) with $\sigma = 0$. Thus, for the standard Bénard problem $R_E \equiv R_L$, i.e., the linear instability boundary $\equiv$ the nonlinear stability boundary, and so no sub-critical instabilities are possible. This result was first established by Joseph (1965,1966).

It is tempting to generalize the natural energy in (3.59) by adding a parameter, λ say, and considering a functional

$$E(t) = \frac{1}{2}\|\mathbf{u}\|^2 + \frac{1}{2}\lambda Pr\|\theta\|^2 \, ,$$

with the idea of employing λ in an optimal way. This is, in fact, precisely what Joseph (1965,1966) did, and his technique is described in the next chapter.

4

Symmetry, Competing Effects, and Coupling Parameters; Multiparameter Eigenvalue Problems; Finite Geometries

4.1 Symmetry, Competing Effects, and Coupling Parameters

In this section we examine two important ideas in the theory of energy stability: the idea of coupling parameters, which was briefly introduced in the last chapter and originally developed by Professor D.D. Joseph, and the concept of symmetry, which is a major reason why the energy method works so well for convection problems.

It is convenient to investigate the above topics in the setting of Bénard convection with an internal heat source, while we also allow the gravity g to depend on the vertical coordinate z. To this end we introduce the relevant equations for a fluid and for fluid flow in a porous solid.

We shall adopt the Boussinesq approximation and so in the body force term we write the density as

$$\rho = \rho_0 \Big[1 - \alpha(T - T_0) \Big], \qquad (4.1)$$

where ρ_0 is a constant, T is temperature, T_0 is a reference temperature and $\alpha \ (= -\rho_0^{-1}(\partial\rho/\partial T)(T_0))$ is the thermal expansion coefficient. For convective motion in an incompressible Newtonian fluid, with spatially varying gravity field and an imposed internal heat source, the relevant equations are then

$$v_{i,t} + v_j v_{i,j} = -\frac{1}{\rho_0} p_{,i} + \nu \Delta v_i - k_i g(z)\big(1 - \alpha[T - T_0]\big), \qquad (4.2)$$

$$v_{i,i} = 0, \qquad (4.3)$$

$$T_{,t} + v_i T_{,i} = \kappa \Delta T + Q(z), \qquad (4.4)$$

where $\mathbf{v}, p, \nu, \Delta, \kappa,$ and Q are, respectively, velocity, pressure, viscosity, the Laplacian operator, thermal diffusivity, and internal heat source, and $\mathbf{k} =$

$(0, 0, 1)$. The analogous equations for convective fluid motion in a porous solid, according to Darcy's law, may be derived as in Joseph (1976b) and may be written (a derivation may also be obtained from the continuum theory for a mixture of an elastic solid and a fluid)

$$\rho_0 A v_{i,t} = -p_{,i} - \frac{\mu}{k} v_i - k_i \rho_0 g(z)\big(1 - \alpha[T - T_0]\big), \tag{4.5}$$

$$v_{i,i} = 0, \tag{4.6}$$

$$T_{,t} + v_i T_{,i} = \kappa \Delta T + Q(z). \tag{4.7}$$

In these equations $\mathbf{v}$ is the seepage velocity, A is an inertia coefficient, μ is dynamic viscosity, k is the permeability of the porous medium, and Q and the time t are adjusted for porosity and specific heat variations between the fluid and solid. Furthermore, it is usual to neglect the left-hand side of (4.5) since the fluid inertia is usually small for flow in porous media; for this chapter we shall do precisely this. Both sets of equations (4.2)–(4.4) and (4.5)–(4.7) are defined on the spatial region $\{z \in (0, d)\} \times \mathbf{R}^2$.

The conduction solution to either set of equations that satisfies the boundary conditions

$$T = T_\ell, \quad z = 0; \qquad T = T_u, \quad z = d; \tag{4.8}$$

where $T_\ell > T_u$, is

$$\bar{\mathbf{v}} \equiv \mathbf{0}, \qquad \bar{T} = -\frac{1}{\kappa} \int_0^z \int_0^\xi Q(\eta) d\eta\, d\xi - cz + T_\ell, \tag{4.9}$$

where the constant c is given by

$$c = \frac{T_\ell - T_u}{d} - \frac{1}{\kappa d} \int_0^d \int_0^\xi Q(\eta) d\eta\, d\xi, \tag{4.10}$$

and where the hydrostatic pressure, $\bar{p}(z)$, is determined from either (4.2) or (4.5), whichever is relevant.

To investigate the stability/instability of (4.9) we introduce perturbations (not necessarily small) such that

$$\mathbf{v} = \bar{\mathbf{v}} + \mathbf{u}, \qquad T = \bar{T} + \theta, \qquad p = \bar{p} + p, \tag{4.11}$$

and then calculate the appropriate equations for the perturbations

$$\mathbf{u}(\mathbf{x}, t), \ \theta(\mathbf{x}, t), \ p(\mathbf{x}, t).$$

These equations are non-dimensionalized according to the scalings:

$$\mathbf{x} = \mathbf{x}^* d, \qquad \mathbf{u} = \mathbf{u}^* U, \qquad U = \frac{\nu}{d}, \qquad \theta = \theta^* T^\sharp,$$

$$T^\sharp = \left(\frac{Pr\, c}{g\alpha}\right)^{1/2} U, \qquad P = \frac{U \rho_0 \nu}{d}, \qquad p = p^* P, \qquad Pr = \frac{\nu}{\kappa}, \tag{4.12}$$

$$t^* = \frac{t\nu}{d^2}, \qquad \delta q(z) = \frac{F(z)}{c}, \qquad R = \left(\frac{cg\alpha d^4}{\kappa \nu}\right)^{1/2},$$

for the fluid case (4.2)–(4.4), where the starred quantities are dimensionless, Pr is the Prandtl number, R^2 is the Rayleigh number, $\delta q(z)$ is defined as indicated with

$$F(z) = \frac{1}{\kappa} \int_0^z Q(\xi)d\xi, \qquad (4.13)$$

and δ is a constant being a scaling for $q(z)$.

The non-dimensional perturbation equations for the fluid are then (omitting all stars)

$$u_{i,t} + u_j u_{i,j} = -p_{,i} + H(z)R\theta k_i + \Delta u_i, \qquad (4.14)$$

$$u_{i,i} = 0, \qquad (4.15)$$

$$Pr(\theta_{,t} + u_i\theta_{,i}) = RN(z)w + \Delta\theta, \qquad (4.16)$$

where $\mathbf{u} = (u, v, w)$, $N(z) = 1 + \delta q(z)$ and $H(z) = 1 + \epsilon h(z)$, $g(z)$ having been defined by $g(z) = g[1 + \epsilon h(z)]$, g constant, and ϵ being a scale for h. These equations hold on the region $\{(x, y) \in \mathbf{R}^2\} \times \{z \in (0, 1)\}$ and

$$\theta = 0, \quad z = 0, 1; \qquad \mathbf{u} = \mathbf{0}, \quad z = 0, 1. \qquad (4.17)$$

A word of caution is in order concerning the Rayleigh number. We have tacitly assumed $c > 0$; if this is not the case we employ $|c|$ with an appropriate insertion of the signum function for R. Also, since c involves Q, which defines $q(z)$, care must be taken in interpreting results involving R^2; however, our definition is consistent with that of Joseph & Shir (1966) who studied the equivalent problem for g constant.

An equivalent analysis for the porous medium equations (4.5)–(4.7) commences again with the steady solution (4.9). The scalings change in so far as ν is replaced by μ/ρ_0 in $P, U,$ and Pr. The temperature scale and Rayleigh number are replaced by

$$T^\sharp = U\left(\frac{c\mu d^2}{k\rho_0 g\alpha\kappa}\right)^{1/2}, \qquad R = \left(\frac{d^2 k\rho_0 g\alpha}{\mu\kappa}\right)^{1/2}, \qquad (4.18)$$

the other scalings remaining the same. With this replacement, the perturbation equations arising from (4.5)–(4.7) are (again all variables are dimensionless), if we neglect inertia

$$p_{,i} = H(z)R\theta k_i - u_i, \qquad (4.19)$$

$$u_{i,i} = 0, \qquad (4.20)$$

$$Pr(\theta_{,t} + u_i\theta_{,i}) = RN(z)w + \Delta\theta, \qquad (4.21)$$

with θ satisfying (4.17) and,

$$u_i n_i = \pm w = 0, \quad \text{on} \quad z = 0, 1, \qquad (4.22)$$

with $\mathbf{n}$ being the unit outward normal to the planes $z = 0, 1$, and where the plus sign is taken if $z = 1$, the minus sign when $z = 0$.

To illustrate Joseph's method of coupling parameters we return to the standard Bénard problem with zero internal heat source and constant gravity field. Equations (4.14), (4.16) have now $H \equiv 1$, $N \equiv 1$, i.e., $\epsilon = \delta = 0$. In accordance with the observed cellular instability, we assume u_i, θ, p have a periodic behaviour in x, y consistent with a shape that tiles the plane; typical are hexagons or two-dimensional rolls. The next step is to define an energy $E(t)$ by

$$E(t) = \frac{1}{2}\|\mathbf{u}\|^2 + \frac{1}{2}\lambda Pr\|\theta\|^2 , \qquad (4.23)$$

where λ is a positive constant to be selected and $\|\cdot\|$ denotes the norm on $L^2(V)$, V being the period cell of u_i, θ, p. By differentiating $E(t)$ and using (4.14)–(4.16) and the boundary conditions, with $< \cdot >$ denoting integration over V it is easily shown that

$$\frac{dE}{dt} = RI - \mathcal{D}, \qquad (4.24)$$

where now

$$I = (1 + \lambda) < \theta w >, \qquad (4.25)$$

$$\mathcal{D} = D(\mathbf{u}) + \lambda D(\theta), \qquad (4.26)$$

$D(\cdot)$ denoting the Dirichlet integral.

Next, as in chapter 3, from (4.24) we derive

$$\frac{dE}{dt} \leq -\mathcal{D}R\Big(\frac{1}{R} - \frac{1}{R_E}\Big), \qquad (4.27)$$

where

$$R_E^{-1} = \max_{\mathcal{H}} \frac{I}{\mathcal{D}}, \qquad (4.28)$$

where $\mathcal{H}$ is the space of admissible functions.

If then

$$R < R_E , \qquad (4.29)$$

$R^{-1} - R_E^{-1} = a$ (say) > 0 and since by Poincaré's inequality $\mathcal{D} \geq c_1 E$ for a positive constant c_1, it follows that

$$\frac{dE}{dt} \leq -ac_1 R E , \qquad (4.30)$$

and so

$$E(t) \leq e^{-ac_1 Rt} E(0). \qquad (4.31)$$

Thus, all disturbances decay very rapidly in the sense of (4.31). The nonlinear stability of the (now linear in z) steady solution (4.9), therefore, requires condition (4.29) to be satisfied.

Before finding the maximum in (4.28) we first introduce the variable $\phi = \lambda^{1/2}\theta$, then

$$\frac{I}{D} = \frac{1+\lambda}{\sqrt{\lambda}} \frac{<\phi w>}{D(\mathbf{u}) + D(\phi)}.$$

The Euler-Lagrange equations for the maximum of this quotient (which are subject to the constraint $\operatorname{div} \mathbf{u} = \mathbf{0}$) are

$$\begin{aligned}
2\Delta\phi + R_E M w &= 0, \\
2\Delta u_i + R_E M \phi k_i &= 2\pi_{,i},
\end{aligned} \tag{4.32}$$

where π arises because of the solenoidal character of $\mathbf{u}$ and

$$M = \frac{1+\lambda}{\sqrt{\lambda}}. \tag{4.33}$$

Equations (4.32) are now to be solved for the lowest eigenvalue R_E. It is expedient at this point to investigate the best value of λ.

The best value of λ, $\bar{\lambda}$, satisfies

$$\bar{\lambda} = 1. \tag{4.34}$$

This result is due to Joseph (1965,1966). We include a proof for a more general situation than (4.32), this situation encompassing other more complicated convection problems.

Note that (4.32) is contained in the more general eigenvalue problem of form

$$L(\mathbf{x})u + R_E f(\mathbf{x}; \lambda)u = G, \tag{4.35}$$

where L is a symmetric, linear differential operator on a suitable Hilbert space and $< G, u >= 0$, $<,>$ being the inner product on the associated Hilbert space. Let now u_1, u_2 be the first eigenfunctions to (4.35) corresponding to parameters λ_1, λ_2. Take the inner product of the equation for u_1 with u_2 and the inner product of the equation for u_2 with u_1. Using the symmetry of L leads to

$$R_E(\lambda_2) < f(\lambda_2)u_2, u_1 > -R_E(\lambda_1) < f(\lambda_1)u_1, u_2 >= 0.$$

Next, add and subtract $R_E(\lambda_1) < f(\lambda_2)u_2, u_1 >$, divide by $\lambda_2 - \lambda_1$, and let $\lambda_2 \to \lambda_1$ to obtain

$$\frac{\partial R_E}{\partial \lambda} < f(\lambda)u, u > + R_E \left\langle \frac{\partial f}{\partial \lambda} u, u \right\rangle = 0.$$

The maximum value of $R_E(\lambda)$ satisfies $\partial R_E / \partial \lambda = 0$ and so

$$\left\langle \frac{\partial f}{\partial \lambda} u, u \right\rangle = 0.$$

For equations (4.32) f is determined in terms of M of (4.33). The condition $\partial M/\partial\lambda = 0$ then yields condition (4.34). Thus, Joseph's method of coupling parameters applied to the *standard* Bénard problem shows $\bar{\lambda} = 1$, i.e., the natural energy is the best choice.

Equations (4.32) with $\lambda = 1$ are

$$\Delta u_i + R_E\theta k_i = \pi_{,i}\,,$$
$$u_{i,i} = 0\,,\qquad\qquad\qquad (4.36)$$
$$\Delta\theta + R_E w = 0\,.$$

These equations are, of course, precisely those of (4.14)–(4.16) with the nonlinear terms omitted, with the time derivatives taken out, and with $H \equiv N \equiv 1$. It was shown in chapter 3 that for the standard Bénard problem the growth rate σ (in the representation $\mathbf{u} = e^{\sigma t}\mathbf{u}(\mathbf{x})$, etc.) is real, and so again we find equations (4.36) are exactly the same as those of the linear theory. Thus, the critical Rayleigh number of nonlinear energy theory that guarantees *stability* is exactly the same as the critical Rayleigh number of linear theory that yields *instability*. Therefore, no sub-critical instabilities can arise.

In fact, the key to the coincidence of the critical Rayleigh numbers for energy and linear theory is the fact that the operator attached to the linear theory is symmetric. This evidently was recognised by Joseph, Davis (1971), and independently and in a different context by Galdi & Straughan (1985a). We outline the ideas of the last contribution.

Let $\mathcal{H}$ be a Hilbert space endowed with a scalar product $(,)$ and associated norm $\|\cdot\|$. We consider in $\mathcal{H}$ the following initial-value problem,

$$u_t + Lu + N(u) = 0,\qquad u(0) = u_0.\qquad (4.37)$$

Here L represents a linear operator (possibly unbounded), and N is a nonlinear operator with $N(0) = 0$ in order that (4.37) admits the null solution. We assume:

(i) L is a densely defined closed operator such that $(L - \lambda I)^{-1}$ is compact for some complex number λ (I is the identity operator in $\mathcal{H}$), that is, L is an operator with compact resolvent;

(ii) The bilinear form associated with L is defined (and bounded) on a space $\mathcal{H}^*$, which is compactly embedded in $\mathcal{H}$.

(iii) The nonlinear operator N verifies the condition

$$(N(u), u) \geq 0,\qquad \forall u \in D(N),\qquad (4.38)$$

where $D(\cdot)$ denotes the domain of the associated operator.

Thanks to (i) the following result is true (Kato (1976), pp. 185–187). *The spectrum of the operator L consists entirely of an at most denumerable number of eigenvalues $\{\sigma_n\}_{n\in\mathbf{N}}$ with finite (both algebraic and geometric) multiplicities and, moreover, such eigenvalues can cluster only at infinity.*

Since the operator L is in general non-symmetric the eigenvalues, which satisfy the equation

$$L\phi = \sigma\phi,$$

are not necessarily real; they may, however, be ordered in the following manner:

$$\mathrm{Re}\,(\sigma_1) \leq \mathrm{Re}\,(\sigma_2) \leq \cdots \leq \mathrm{Re}\,(\sigma_n) \leq \cdots . \tag{4.39}$$

In accordance with standard literature on stability theory we include the following definitions.

Definition 1. The zero solution to (4.37) is said to be *linearly stable* if and only if

$$\mathrm{Re}\,(\sigma_1) > 0. \tag{4.40}$$

Definition 2. The zero solution to (4.37) is said to be *nonlinearly stable* if and only if for each $\epsilon > 0$ there is a $\delta = \delta(\epsilon)$ such that

$$\|u_0\| < \delta \quad \Rightarrow \quad \|u(t)\| < \epsilon, \tag{a}$$

and there exists γ with $0 < \gamma \leq \infty$ such that

$$\|u_0\| < \gamma \quad \Rightarrow \quad \lim_{t\to\infty} \|u(t)\| = 0. \tag{b}$$

If $\gamma = \infty$, we say the zero solution is unconditionally nonlinearly stable (or simply refer to it as being nonlinearly asymptotically stable), otherwise for $\gamma < \infty$ the solution is conditionally stable. The value of γ is called the size of the *attracting radius*.

The relationship between linear and nonlinear stability is now investigated.

The operator L is in general non-symmetric, although it allows a decomposition into two parts L_1 and L_2 such that
(i) $L = L_1 + L_2$, $D(L_2) \supset D(L_1) = D(L)$;
(ii) L_1 is symmetric, with compact resolvent;
(iii) L_2 is skew-symmetric and bounded in $\mathcal{H}^*$.
From (ii) it follows that the spectrum of L_1 satisfies the same type of result as that given above for the spectrum of L. Moreover, because of the symmetry, the eigenvalues $\{\lambda_n\}_{n\in\mathbf{N}}$ associated with L_1 are *all* real and may be ordered

$$\lambda_1 \leq \lambda_2 \leq \cdots \leq \lambda_n \leq \cdots .$$

Let $L_1[\phi, \phi]$, $\phi \in \mathcal{H}^*$, be the bilinear form associated with the operator L_1, i.e.,

$$(L_1\phi, \phi) = L_1[\phi, \phi], \qquad \forall\phi \in D(L_1).$$

Under the above conditions, the following lemma holds.

Lemma 4.1.
Let $\bar{\phi}$ be a (normalized) eigenfunction associated with the eigenvalue λ_1.
Then

$$\lambda_1 = L_1[\bar{\phi}, \bar{\phi}] = \min_{\phi \in \mathcal{H}^*} \frac{L_1[\phi, \phi]}{\|\phi\|^2}.$$

The following result establishes unconditional nonlinear stability.
Suppose

$$\lambda_1 > 0. \tag{4.41}$$

Then the zero solution to (4.37) is unconditionally nonlinearly stable.
Proof.
Form the scalar product of (4.37) with u to obtain

$$\frac{1}{2}\frac{d}{dt}\|u\|^2 + (Lu, u) + (N(u), u) = 0. \tag{4.42}$$

Since L_2 is skew-symmetric and since by (4.38) N is non-negative, there
follows

$$\frac{1}{2}\frac{d}{dt}\|u\|^2 + \frac{L_1[u, u]}{\|u\|^2}\|u\|^2 \le 0.$$

With the aid of Lemma 4.1 we thus derive

$$\frac{1}{2}\frac{d}{dt}\|u\|^2 + \lambda_1\|u\|^2 \le 0, \tag{4.43}$$

and so

$$\|u(t)\|^2 \le \|u_0\|^2 e^{-2\lambda_1 t}.$$

In the light of (4.41), the result follows at once.

From the above arguments it follows that while the linear stability problem is reduced to studying the eigenvalue problem associated with all of
L, nonlinear stability according to the standard energy method described
in the above result involves the eigenvalues of the *symmetric* part L_1 only.
Moreover, whenever $L_2 = 0$, the two eigenvalue problems coincide and *linear stability always implies nonlinear stability.* (The converse is also true.)

Even though Galdi & Straughan (1985a) show that convection between
spherical shells with the gravitational potential proportional to the steady
temperature distribution results in a symmetric linearized problem, and
magnetohydrodynamic convection in a "quasi-static" electric field approximation with zero limit of magnetic Prandtl number also enjoys such a
property, symmetry is too strong a condition to expect in general. Indeed, Busse (1967) uses weakly nonlinear analysis to show that sub-critical
bifurcation will occur if either the fluid viscosity or expansion coefficient
depends on temperature. In reality fluid properties are not constant and
so the linearized system will not, in general, be symmetric. Fortunately,
however, for what would appear to be the majority of convection problems,

the non-symmetric part acts like a bounded operator when compared to
the symmetric (unbounded) operator. This fact is closely connected to the
reason why the energy method often yields nonlinear stability results very
close to those of linear theory, thereby yielding a small band of Rayleigh
numbers where possible sub-critical instability may arise. In such a case
energy theory is also very useful in that it shows the linear theory has
essentially captured the physics of the onset of convection.

In precisely the situation outlined above Davis (1969a) established an interesting result. It roughly says that if one adds a bounded (non-symmetric)
operator to (4.37) then the eigenvalue of linear theory remains real provided
the non-symmetric operator is small enough. More precisely Davis (1969a)
considers the reduced eigenvalue problem

$$Au_0 = \sigma_0 B u_0 \,, \tag{4.44}$$

with associated homogeneous boundary conditions where
  (i) u_0 is an N-component vector of functions;
  (ii) A is a selfadjoint linear operator mapping real vectors into real vectors;
  (iii) B is a real $N \times N$ constant, diagonal matrix $\mathrm{diag}(b_i)$, with
$0 < b_i < \infty$ $(i = 1, ..., N)$;
  (iv) (4.44) has a discrete spectrum $\{\sigma_0^{(n)}\}$ each $\sigma_0^{(n)}$ having at most finite
multiplicity and ordered so that

$$|\sigma_0^{(1)}| \le |\sigma_0^{(2)}| \le \cdots \le |\sigma_0^{(k)}| \le \cdots$$

and corresponding eigenfunctions $\{u_{0n}\}$ that form a complete set under the
inner product (u_{0n}, Bu_{0n}) where $(u_{0k}, u_{0k})^{1/2}$ is the L^2 norm of u_{0k}.

He compares this to the perturbed eigenvalue problem

$$Au - \epsilon Mu = \sigma Bu \qquad (\epsilon > 0), \tag{4.45}$$

with associated homogeneous boundary conditions where
  (v) M is an operator of L^2 norm one,

$$\|M\| = \sup\{\|Mx\| \,\big|\, \|x\| \le 1\} = 1,$$

which maps real vectors into real vectors and Mu belongs to the complete
space of (iv).

By writing the solution to (4.45) as a power series in ϵ and finding a
radius of convergence for this power series, Davis (1969a, p. 343) shows
that
The kth eigenvalue $\sigma^{(k)}$ of the perturbed problem (4.45) for which

$$\lim_{\epsilon \to 0} \sigma^{(k)} = \sigma_0^{(k)}$$

is real, provided that $\sigma_0^{(k)}$ *is simple and*

$$\epsilon < \epsilon_c^{(k)} = \mathcal{Q}\mathcal{L}_k^{-1} \qquad (k = 1, 2, ...), \qquad (4.46)$$

where

$$\mathcal{Q} = B_m \left[1 + 2B_M B_m^{-1} - 2(B_M^2 B_m^{-2} + B_M B_m^{-1})^{1/2}\right], \qquad (4.47)$$

$$B_M = \max\{b_1, ..., b_N\}, \qquad B_m = \min\{b_1, ..., b_N\}, \qquad (4.48)$$

and $\mathcal{L}_k^{-1}$ *is the minimum non-zero value of* $|\sigma_0^{(k)} - \sigma_0^{(n)}|$, *minimized over* n, $n \neq k$.

While this is certainly an interesting result, its practical use is limited largely due to the need to calculate $\mathcal{L}_k^{-1}$. Nevertheless, it does show there is a range of ϵ for which the first eigenvalue of linear theory remains real. Davis (1969a) calculates ϵ_c for several problems, of especial interest here being his application to non-Boussinesq penetrative convection. Unfortunately, he remarks (p. 353) that for the model of Veronis (1963) his study, *sheds no light on the (above) situation since the radius of convergence bound is too small to cover situations where* ρ/ρ_0 *has a maximum in the interior of the layer.*

We now return to (4.14)–(4.16), the linearized form of which satisfy Davis' theorem for ϵ, δ small enough. With $\epsilon = 0$, nonlinear energy stability of (4.14)–(4.16) has been studied in detail by Joseph & Shir (1966) while a nonlinear energy analysis for (4.19)–(4.21) with $\epsilon = 0$ is contained in Joseph (1976b). On the other hand, a nonlinear energy stability analysis for (4.14)–(4.16) with $\delta = 0$ is given by Straughan (1989). These problems, i.e., when $\epsilon = 0$ or $\delta = 0$ but not both zero are very special in that exchange of stabilities holds, at least for two *stress free* boundaries. This follows from E.A. Spiegel's method, given in Veronis (1963). (If we write $u(\mathbf{x}, t) = e^{\sigma t}\phi(\mathbf{x})$ in the linearized version of (4.37) then ϕ satisfies

$$\sigma\phi + L\phi = 0,$$

where, in general, $\sigma = \sigma_r + i\sigma_1$, $\sigma_r, \sigma_1 \in \mathbf{R}$: in the present context exchange of stabilities is said to hold if $\sigma_1 \neq 0 \Rightarrow \sigma_r < 0$.) We outline the idea of Spiegel's method here.

Suppose $\delta = 0$ in (4.16), linearize by removing the convective terms from (4.14) and (4.16) and assume a solution like $\mathbf{u} = e^{\sigma t}\mathbf{u}(\mathbf{x})$ with a similar representation for θ and p. Next, take curlcurl of the equation arising from (4.14) and take the third component of the resulting equation to obtain the system

$$\sigma\Delta w = \Delta^2 w + RH(z)\Delta^*\theta, \qquad (4.49)$$

$$\sigma Pr\theta = \Delta\theta + Rw, \qquad (4.50)$$

where $\Delta^* = \partial^2/\partial x^2 + \partial^2/\partial y^2$. For two free boundaries the relevant boundary conditions are

$$w = \Delta w = \theta = 0, \qquad z = 0, 1. \tag{4.51}$$

The variable w is eliminated from (4.49), (4.50) to yield

$$\Delta^3\theta - \sigma(1 + Pr)\Delta^2\theta + \sigma^2 Pr\Delta\theta - R^2 H(z)\Delta^*\theta = 0. \tag{4.52}$$

Multiply (4.52) by θ^* (complex conjugate of θ) and integrate over a cell, V, of solution periodicity. With $\|\cdot\|$ and $< \cdot >$ again denoting the norm on $L^2(V)$ and integration over V, respectively, we find

$$-\|\nabla\Delta\theta\|^2 - \sigma(1 + Pr)\|\Delta\theta\|^2 - \sigma^2 Pr\|\nabla\theta\|^2 + R^2 < H|\nabla^*\theta|^2 > = 0,$$

where $\nabla^* = (\partial/\partial x, \partial/\partial y)$. The imaginary part of this equation leads to

$$\sigma_1\left[(1 + Pr)\|\Delta\theta\|^2 + 2\sigma_r Pr\|\nabla\theta\|^2\right] = 0,$$

where it has been assumed that $\sigma = \sigma_r + i\sigma_1$. Hence, if $\sigma_1 \neq 0$ then $\sigma_r < 0$. This shows it is sufficient to consider the stationary convection boundary $\sigma \equiv 0$.

The same conclusion is easily arrived at for (4.14)–(4.16) when $\epsilon = 0$, $\delta \neq 0$. Here, instead of eliminating w, θ is eliminated to obtain instead of (4.52), a sixth order equation in w. The rest of the proof is similar.

It would appear that the importance of this result for energy stability is that the linear and nonlinear boundaries are very close, even when δ or ϵ is large (with one of them zero), thus rendering the nonlinear results very useful, even for situations where Davis' theorem does not apply. At the time of writing I have not been able to establish a result of the above type when $\epsilon \neq 0$, $\delta \neq 0$ for either the linearized form of (4.14)–(4.16) or the seemingly simpler linearized form of (4.19)–(4.21). The above proof seems to need the $H(z)$ or $N(z)$ term to be multiplied by derivatives not involving z, i.e., Δ^*. This, of course, raises the interesting possibility that σ might be complex at criticality and overstable convection the dominant mode of instability, at least for some range of ϵ, δ. However, if $\epsilon h \equiv \delta q \neq 0$, the linear system is symmetric and so $\sigma \in \mathbf{R}$: also numerical work has so far indicated stationary convection is the dominant mode.

This section is completed with an energy analysis of (4.19)–(4.21). The separate energy identities are derived by multiplying (4.19) by u_i, (4.21) by θ and integrating over V to obtain

$$\frac{1}{2}Pr\frac{d}{dt}\|\theta\|^2 = R < N\theta w > -D(\theta), \tag{4.53}$$

$$R < Hw\theta > = \|\mathbf{u}\|^2. \tag{4.54}$$

Hence, for $\lambda(> 0)$ to be chosen we may form the *energy* identity,

$$\frac{dE}{dt} = R < M(z)w\theta > -D(\theta) - \lambda\|\mathbf{u}\|^2, \tag{4.55}$$

where

$$E(t) = \frac{1}{2}Pr\|\theta\|^2, \tag{4.56}$$

$$M(z) = 1 + \delta q + \lambda(1 + \epsilon h). \tag{4.57}$$

If we define R_E by

$$R_E^{-1} = \max_{\mathcal{H}} \frac{I}{\mathcal{D}} \tag{4.58}$$

with

$$I = < Mw\theta >, \qquad \mathcal{D} = D(\theta) + \lambda\|\mathbf{u}\|^2,$$

then as before it is straightforward to show $E(t) \to 0$ at least exponentially provided $R < R_E$.

It is expedient to put $\mathbf{v} = \lambda^{1/2}\mathbf{u}$ in (4.58), then with

$$F(z) = \frac{M(z)}{\lambda^{1/2}} \tag{4.59}$$

and setting $v_3 = w$, the Euler-Lagrange equations arising from (4.58), for the determination of R_E, are

$$\Delta\theta + \frac{1}{2}FR_Ew = 0,$$
$$v_i - \frac{1}{2}FR_E\theta k_i = \pi_{,i}, \tag{4.60}$$

v_i again being solenoidal.

In this case by employing parametric differentiation we may show that with ξ standing for δ, ϵ, or λ,

$$R_E^2\left\langle \frac{\partial F}{\partial \xi} w\theta \right\rangle + \mathcal{D}\frac{\partial R_E}{\partial \xi} = 0, \tag{4.61}$$

where

$$\mathcal{D} = \|\mathbf{v}\|^2 + D(\theta).$$

At the best value of λ, $\partial R_E/\partial\lambda = 0$ and then (4.61) gives rise to

$$\left\langle w\theta\{\bar{\lambda}(1 + \epsilon h) - (1 + \delta q)\}\right\rangle = 0.$$

This suggests that a good "guess" (useful when searching numerically) is

$$\tilde{\lambda} = \frac{1 + \delta q_a}{1 + \epsilon h_a},$$

where q_a, h_a denote average values. If $h = -z$, $q = z$, then

$$\tilde{\lambda} = \frac{1 + \frac{1}{2}\delta}{1 - \frac{1}{2}\epsilon}. \tag{4.62}$$

Taking ξ to be ϵ or δ in (4.61) yields

$$
\begin{aligned}
\mathcal{D}\frac{\partial R_E}{\partial \epsilon} &= -\lambda^{1/2} R_E^2 < hw\theta >, \\
\mathcal{D}\frac{\partial R_E}{\partial \delta} &= -\lambda^{-1/2} R_E^2 < qw\theta > .
\end{aligned}
\tag{4.63}
$$

Experience with numerical eigenvalue problems of this type suggests terms like $< hw\theta >$ are frequently one signed and thus (4.63) are useful guides to the qualitative behaviour of R_E with ϵ or δ.

Since F depends on z, (4.60) is solved numerically. The periodicity is exploited to look for an x and y dependence with

$$\Delta^*\theta = -a^2\theta, \tag{4.64}$$

where a is a wavenumber. After removing π by operating by curlcurl equations (4.60) may then be reduced to

$$
\begin{aligned}
D^2 W &= a^2 W - \frac{1}{2} R_E F a^2 \Theta, \\
D^2 \Theta &= a^2 \Theta - \frac{1}{2} R_E F W,
\end{aligned}
\tag{4.65}
$$

where $w = W(z)w(x, y)$, $\theta = \Theta(z)\theta(x, y)$, the x, y parts satisfying a relation like (4.64), and $D = d/dz$.

The boundary conditions are

$$W = \Theta = 0 \quad \text{on} \quad z = 0, 1. \tag{4.66}$$

System (4.65), (4.66) constitutes a fourth order eigenvalue problem with non-constant coefficients. It was solved numerically by the compound matrix method (see Appendix 2). Since $x, y \in \mathbf{R}^2$, we seek the most unstable wavenumber. On the other hand, the "energy" parameter λ may be chosen to make R_E as large as possible. Hence we must solve the numerical problem

$$\max_{\lambda} \min_{a^2} R_E(a^2; \lambda). \tag{4.67}$$

The derivatives in the optimization problem are not known and so a quasi-Newton (or similar) technique that does not require derivatives would have to be used. I have found the golden section search method works well on both the maximum and minimum problems. It may be a little more expensive in computer time than other techniques, but for problems like (4.67) it has been very reliable.

The following numerical results are a selection taken from many presented by Rionero & Straughan (1990). In Tables 4.1 to 4.3 we present the critical Rayleigh and wavenumbers for energy and linear theory. In these tables, subscripts E and L denote energy and linear values, respectively, and $\bar{\lambda}$ is the best value of λ.

R_L^2	R_E^2	a_L^2	a_E^2	ϵ	δ	$\bar{\lambda}$
39.478	39.478	9.870	9.870	0.0	0.0	1.000
77.020	74.971	10.209	10.178	1.0	0.0	1.500
132.020	122.374	12.314	11.928	1.5	0.0	2.565
63.114	61.563	10.026	10.110	1.0	0.5	2.318
111.930	101.513	11.887	11.562	1.5	0.5	3.201
35.384	35.092	9.835	9.870	0.5	1.0	2.000
53.375	51.715	9.918	10.032	1.0	1.0	2.797
165.748	137.350	18.612	14.853	1.9	1.0	4.268

Table 4.1. Critical Rayleigh and wave numbers for
$h = -z$, $q = z$. ($\bar{\lambda}$ denotes the best value of λ in (4.67).)
(After Rionero & Straughan (1990).)

R_L^2	R_E^2	a_L^2	a_E^2	ϵ	δ	$\bar{\lambda}$
59.053	58.648	10.005	10.028	0.5	0.0	1.461
31.552	31.528	9.884	9.887	0.0	0.5	1.254
47.130	47.017	9.890	9.899	0.3	0.1	1.306
36.848	36.815	9.867	9.870	0.1	0.3	1.234
43.154	42.997	9.869	9.884	0.3	0.3	1.432

Table 4.2. Critical Rayleigh and wave numbers for
$h = -(e^z - 1)$, $q = z$. ($\bar{\lambda}$ denotes the best value of λ in (4.67).)
(After Rionero & Straughan (1990).)

R_L^2	R_E^2	a_L^2	a_E^2	ϵ	δ	$\bar{\lambda}$
54.390	53.943	10.034	10.058	1.0	0.0	1.341
36.549	36.498	9.892	9.899	0.0	1.0	1.085
46.760	46.613	9.898	9.910	0.6	0.2	1.210
40.009	39.955	9.865	9.871	0.2	0.6	1.111
45.538	45.312	9.870	9.889	0.6	0.6	1.249

Table 4.3. Critical Rayleigh and wave numbers for
$h = -z^2$, $q = z^5$. ($\bar{\lambda}$ denotes the best value of λ in (4.67).)
(After Rionero & Straughan (1990).)

One conclusion immediately evident from these tables is that the energy values R_E^2 are very close to the linear values R_L^2. Even for the worst case where the gravity changes direction the values only differ by approximately 17% (the case $\epsilon = 1.9$, $\delta = 1.0$ in Table 4.1).

For completeness we include in Table 4.4 the values of the complex eigenvalue σ for the problem covered in Table 4.1, with $\epsilon = 1.9$, $\delta = 1.0$. Even though the energy limit is approximately 17% below the linear one, the linear eigenvalue is real at the onset of convection. To obtain the results in Table 4.4 we fixed $Ra = R_L^2$ and a_L^2, and calculated $\sigma = \sigma_r + i\sigma_1$ using the compound matrix method.

Ra	a_L^2	σ_r	σ_1
165.748	18.612	1.051×10^{-4}	2.814×10^{-16}
165.740	18.600	-1.694×10^{-3}	3.118×10^{-16}
165.748	18.600	-1.022×10^{-4}	2.817×10^{-16}
165.760	18.600	2.796×10^{-3}	1.983×10^{-16}
165.760	18.612	2.800×10^{-3}	1.980×10^{-16}
165.760	18.620	2.799×10^{-3}	1.979×10^{-16}
165.748	18.620	1.036×10^{-4}	2.814×10^{-16}
165.740	18.620	-1.694×10^{-3}	3.115×10^{-16}
165.740	18.612	-1.692×10^{-3}	3.116×10^{-16}

Table 4.4. Critical Rayleigh number, wave numbers, and
growth rate: $h = -z$, $q = z$; $\epsilon = 1.9$, $\delta = 1.0$.
(After Rionero & Straughan (1990).)

4.2 Multiparameter Eigenvalue Problems and Their Occurrence in Energy Stability Theory

If there are more than two fields such as velocity and temperature in a convection problem, say a salt field as a third, then it is natural to use an energy that involves more than one coupling parameter. The only works on energy theory I am aware of that use more than one coupling parameter use clever ad hoc tricks to essentially reduce the problem to a one parameter one, thereby just having a numerical calculation like (4.67). A good case in point is Joseph's (1970) analysis of double-diffusive convection.

In general, however, the problem of convection when there are competing temperature and one or more concentration fields may yield sharp results if a more sophisticated numerical approach were used. From the mathematical and physical viewpoint, problems such as thermohaline convection are interesting; they definitely exhibit sub-critical bifurcation, see Veronis (1965), Proctor (1981), and, therefore, a global nonlinear stability threshold is always desirable.

To illustrate the point we might, for example, consider convection with a temperature and a concentration field in the presence of a Soret effect. Such an effect is likely to have importance in semiconductor crystal growth, see Hurle & Jakeman (1971), who consider a nonlinear Soret effect. (The Soret effect is that whereby mass diffusion is influenced by a temperature gradient, i.e., manifested by the $\kappa_S \Delta T$ term in (4.70) below.) If the Soret effect is only *linear* the equations are

$$\dot{v}_i = -\frac{1}{\rho_0} p_{,i} + \nu \Delta v_i - g k_i \left(1 - \alpha[T - T_0] + \gamma[C - C_0]\right), \quad (4.68)$$

$$\dot{T} = \kappa \Delta T, \quad (4.69)$$

$$\dot{C} = \kappa_C \Delta C + \kappa_S \Delta T, \quad (4.70)$$

$$v_{i,i} = 0. \quad (4.71)$$

It is assumed these equations occupy the plane layer $z \in (0, d)$, C is the concentration field, $C_0, \gamma, \kappa_C, \kappa_S$ are constants, and a superposed dot denotes the convective derivative $\partial/\partial t + v_i \partial/\partial x_i$.

Under prescribed boundary conditions on T, C a steady solution linear in z is possible, and the perturbation equations from this solution will be

$$\begin{aligned}
\dot{u}_i &= -p_{,i} + \Delta u_i + R\theta k_i + H R_S \phi k_i, \\
u_{i,i} &= 0, \\
Pr\dot{\theta} &= \hat{H} R w + \Delta \theta, \\
Pc\dot{\phi} &= \tilde{H} R_S w + \Delta \phi + \xi \Delta \theta,
\end{aligned} \quad (4.72)$$

where a non-dimensionalization has been employed, R_S is a Rayleigh number associated with the concentration field, $H, \hat{H}, \tilde{H}$ are ± 1 according to the steady solution (which depends on the boundary conditions), Pc is a concentration Prandtl number, ξ is a constant, and ϕ is the perturbation to C. We assume $\kappa_S \leq 2(\kappa\kappa_C)^{1/2}$ to ensure $\xi \leq 2$.

To study the nonlinear stability of this system we may follow Shir & Joseph (1968) who studied this problem with $\xi = 0$ and introduce an "energy" $E(t)$ of the form

$$E(t) = \frac{1}{2}\|\mathbf{u}\|^2 + \frac{1}{2}Pr\lambda_1\|\theta\|^2 + \frac{1}{2}Pc\lambda_2\|\phi\|^2, \qquad (4.73)$$

where λ_1 and λ_2 are positive coupling parameters to be selected optimally to give as sharp a stability boundary as possible. The energy equation that arises has the form

$$\frac{dE}{dt} = RI - \mathcal{D}, \qquad (4.74)$$

where we have put $R_S = \alpha R$ and

$$I = (1 + \hat{H}\lambda_1) < \theta w > + \alpha(H + \tilde{H}\lambda_2) < \phi w >, \qquad (4.75)$$

$$\mathcal{D} = D(\mathbf{u}) + \lambda_1 D(\theta) + \lambda_2 \big[D(\phi) + \xi D(\theta, \phi)\big], \qquad (4.76)$$

where

$$D(\theta, \phi) = < \nabla\theta.\nabla\phi > .$$

Again, for nonlinear stability it may be demonstrated from (4.74) that $R < R_E$ is a sufficient condition, where R_E is defined by

$$R_E^{-1} = \max_{\mathcal{H}} \frac{I}{\mathcal{D}}. \qquad (4.77)$$

The five Euler-Lagrange equations that arise from this maximum problem together with the solenoidal constraint on $\mathbf{u}$ define a three-parameter eigenvalue problem for the lowest eigenvalue $R_E(a^2; \lambda_1, \lambda_2)$, a being again a wavenumber. The critical value of the nonlinear stability Rayleigh number in this case is $Ra_E = R_E^2$, where

$$R_E = \max_{\lambda_1, \lambda_2} \min_{a^2} R_E(a^2; \lambda_1, \lambda_2). \qquad (4.78)$$

I am unaware of any numerical calculations for (4.78), however, it is clear that a three-way sweep using the golden section search algorithm would not be satisfactory, it would certainly consume a large amount of CPU time. Instead, one would have to use a quasi-Newton technique on the maximization problem, such as Broyden's method described e.g., in Dennis & Schnabel (1983). This is certainly a promising area for tackling energy stability problems when there are many equations and as it also seems that the maximization/minimization routines will be well suited to be spread over the processors in a parallel array of computers, further work should prove rewarding.

The paper of Rionero & Mulone (1987) contains some interesting results concerning a linearization principle for the coincidence of the linear and nonlinear stability boundaries in certain situations, for a system like (4.72), but which contains both linear Soret and Dufour effects. (A Dufour effect essentially adds a term like $\zeta\Delta\phi$ to the right-hand side of (4.72)$_3$.)

4.3 Finite Geometries, Spatially Varying Boundary Conditions, and Numerically Determined Basic Solutions

Up to this point we have described convection problems in which the fluid occupies an infinite plane layer. For many real situations this is inadequate. When, however, we abandon the infinite layer and examine finite regions, usually the method of normal modes is not sufficient, i.e., to resolve the eigenvalue problem of energy theory, it is no longer sufficient to solve an ordinary differential equation, and more sophisticated numerical techniques must be employed. Also, it is quite likely that even the basic solution will have to be determined numerically, although this depends on the boundary conditions. Having to numerically determine the basic solution in the region and then use this in the energy eigenvalue analysis clearly gives a more complicated numerical task.

A fundamental step in the lines of numerically oriented energy analyses was made by Munson & Joseph (1971) in their work on the stability problem of flow between two spheres: this is well documented by Joseph (1976a).

In the field of convection the work of Jankowski et al. (1988) on a crystal growth problem employs energy theory where the basic state is numerically determined. At present their results yield conservative Rayleigh numbers.

Another interesting recent contribution is the paper of Reddy & Voyé (1988). They write out the general equations for convection in an arbitrary shaped finite region and develop an energy analysis for a general base flow, which could be numerically determined. They first show, by using a Galerkin (finite element) penalty technique, that the energy stability problem is well posed. In particular, they show the existence of a maximizing solution to the problem for $R_E(\lambda)$ for a fixed coupling parameter λ. The first existence proof of a maximizing solution was given by Rionero (1967,1968); Rionero's fundamental result (which was published in Italian) is further discussed by Galdi & Straughan (1985a). Reddy & Voyé (1988) point out that it is not yet known whether the value of λ that yields a maximum in $R_E(\lambda)$ is positive. From a practical point of view one usually solves for R_E with λ fixed and then treats the maximum problem as a constrained one. I have seen little trouble with my own numerical results. Nevertheless,

Reddy & Voyé's point is correct and care must be taken. Indeed, in the next chapter we draw attention to the fact that when the region is infinite in all directions the maximizing solution may not exist for all λ.

Reddy & Voyé (1988) do quote a convergence result for their penalty approximation to the energy eigenvalue problem. They also illustrate their results by examining the two-dimensional Bénard convection problem with an internal heat source. Their results are certainly sharp and very detailed. They include several graphs of critical Rayleigh number against the aspect ratio (ℓ/d = length/depth of the two-dimensional rectangular layer), and these are revealing. They indicate that when $\ell/d \approx 8$ the results are nearly identical to those obtained by normal modes for an infinite layer. Also interesting are the results on the number of convection cells at the boundary of stability. For their energy results they include graphs of R_E against λ for $\ell/d = 10$, and these clearly indicate a maximum. No indication of what maximizing technique was used seems to be included.

Certainly the results of Jankowski et al. (1988) and Reddy & Voyé (1988) are an important step toward a nonlinear energy stability theory for general domains. As Reddy & Voyé (1988) point out, however, they have not yet tackled three-dimensional regions. While I believe the energy method is an invaluable tool it loses a lot of its attractive simplicity when the calculations reduce to three-dimensional numerical eigenvalue problems. Indeed, one must then ask whether energy theory will be as cost effective as a direct numerical simulation on the three-dimensional convection problem.

5

Convection Problems in a Half-Space

5.1 Existence in the Energy Maximum Problem

It is often useful to define a convection problem on a half space. For example, Hurle, Jakeman & Wheeler (1982) use the velocity in a phase change problem to transform their stability analysis to one on a half-space; also Hurle, Jakeman & Pike (1967) have heat conducting half-spaces bounding a fluid layer to investigate the effects of finite conductivity at the boundary. While it may offer some simplicity to deal with a half-space configuration, from the mathematical point of view it does introduce new complications. In particular, Galdi & Rionero (1985) derive a very sharp result on the asymptotic behaviour of the base solution for which the energy maximum problem for R_E admits a maximizing solution. Roughly speaking, either the base solution must decay at least linearly at infinity, or the gradient of the base solution must decay at least like $1/z^2$ (if $z > 0$ is the half-space.) To describe this result and related ones in geophysics it is convenient to return to the general equations for a heat conducting linearly viscous fluid.

Let now $V_i(\mathbf{x}), P(\mathbf{x}), T(\mathbf{x})$ be a steady solution to (3.39), with g constant, on the unbounded domain $\Omega = \{z > 0\} \times \mathbf{R}^2$. We suppose V_i, T are prescribed on $z = 0$ and asymptotic behaviour is prescribed as $x, y, z \to \infty$. If we introduce perturbations $u_i(\mathbf{x}, t), p(\mathbf{x}, t), \theta(\mathbf{x}, t)$ to this solution and non-dimensionalize with length and time scales $d, d^2/\nu$, then select base velocity and temperature scales $V, dT'(0)$, where d is some length; select perturbation velocity and temperature scales $U = \nu/d$ and $U(T'(0)\nu/\kappa g\alpha)^{1/2}$, where $T'(0) = dT/dz(0)$; and define the Rayleigh number, $Ra = R^2$, Reynolds number, Re, and Prandtl number, Pr, as

$$R^2 = \frac{T'(0)\alpha g d^4}{\kappa \nu}, \qquad Re = \frac{Vd}{\nu}, \qquad Pr = \frac{\nu}{\kappa}; \qquad (5.1)$$

then the non-dimensional perturbation equations are

$$u_{i,t} + Re(V_j u_{i,j} + V_{i,j} u_j) + u_j u_{i,j} = -p_{,i} + \Delta u_i + R\theta k_i,$$
$$u_{i,i} = 0, \qquad (5.2)$$
$$Pr(\theta_{,t} + u_i \theta_{,i}) + Pr Re V_i \theta_{,i} = -RT_{,i} u_i + \Delta\theta,$$

holding on the region $\Omega = \{z > 0\} \times \mathbf{R}^2$. We now briefly describe theorem 2.3 of Galdi & Rionero (1985); however, they only establish a result for the Navier-Stokes equations and so we here interpret their result for Bénard convection. Furthermore, we stress that Galdi & Rionero (1985) establish existence of a suitable steady solution and rigorously establish existence of the energy maximizing function and a suitable energy theory. Here we only include sufficient technical details to understand the difficulty imposed by the half-space aspect of the problem. Extensions of the results of Galdi & Rionero (1985) to when the basic solution has stronger behaviour at infinity are given by de Angelis (1990): given the stronger behaviour imposed, she obtains better stability criteria.

We can assume periodicity in the (x, y) directions, and let V denote $\{z > 0\} \times$ {a period cell}. We then multiply $(5.2)_1$ by u_i, $(5.2)_3$ by θ and integrate over V. With u_i, θ vanishing on $z = 0$ and decaying sufficiently rapidly at infinity in the sense that

$$\left\langle \frac{u_i u_i + \theta^2}{(1+z)^2} \right\rangle + D(\mathbf{u}) + D(\theta) < \infty, \tag{5.3}$$

where $< \cdot >$ denotes integration over V, $D(\cdot)$ the Dirichlet integral over V, we formally obtain the equations

$$\frac{1}{2}\frac{d}{dt}\|\mathbf{u}\|^2 = -D(\mathbf{u}) + R < \theta w > -Re < D_{ij}u_i u_j >, \tag{5.4}$$

$$\frac{1}{2}Pr\frac{d}{dt}\|\theta\|^2 = -D(\theta) - R < T_{,i}u_i \theta >, \tag{5.5}$$

where $D_{ij} = \frac{1}{2}(V_{i,j} + V_{j,i})$.

Of course, it is *not* clear that these equations make sense since the

$$< \theta w >, \quad < D_{ij}u_i u_j >, \quad < T_{,i}u_i \theta >$$

terms do not in general satisfy (5.3). It is precisely this point that needs great care. The $< \theta w >$ term requires $\theta, w \in L^2(V)$. Even then, we cannot dominate this term by the $D(\theta)$ term since the form of Poincaré's inequality that holds for the half-space has the form (Galdi & Rionero (1985), lemma 2.1)

$$\left\langle \frac{u_i u_i}{(1+z)^2} \right\rangle \leq 4D(\mathbf{u}) \quad \text{and} \quad \left\langle \frac{\theta^2}{(1+z)^2} \right\rangle \leq 4D(\theta). \tag{5.6}$$

Even a decay behaviour at infinity in dT/dz like z^{-2} is not sufficient, at least from (5.4), (5.5). One solution to what is needed is after a finite $z-$length $T_{,i} \sim \beta$ (positive constant), i.e., the *temperature gradient is destabilizing only for a finite $z-$length and after that it stabilizes linearly*. In this way, the

$$R < \theta w > -\lambda R < T_{,i}u_i \theta > \tag{5.7}$$

terms, which will arise in a coupling parameter energy formulation, may be reduced to integrals over a finite region. *Assuming this to be so* we form

$$\frac{d}{dt}\left(\frac{1}{2}\|\mathbf{u}\|^2 + \frac{1}{2}\lambda Pr\|\theta\|^2\right) = -\left[D(\mathbf{u}) + \lambda D(\theta)\right] + R < \theta w >$$
$$- R\lambda < T_{,i}u_i\theta > -Re < D_{ij}u_iu_j > . \tag{5.8}$$

Note that λ has already been selected to remove the troublesome terms in (5.7). The maximum problem is then to determine

$$\max_{\mathcal{H}}\left[\frac{-\alpha < D_{ij}u_iu_j > + < \theta w > -\lambda < T_{,i}u_i\theta >}{D(\mathbf{u}) + \lambda D(\theta)}\right], \tag{5.9}$$

where α is a parameter such that $Re = \alpha R$. The θ terms in the numerator are bounded and to ensure the $D_{ij}u_iu_j$ term is likewise well behaved, Galdi & Rionero (1985) require that either

$$|D_{ij}| = O\left(\frac{1}{(1+z)^2}\right), \qquad \text{as } z \to \infty \tag{5.10}$$

or

$$|\mathbf{V} - \mathbf{V}_\infty| = O\left(\frac{1}{(1+z)}\right), \qquad \text{as } z \to \infty, \tag{5.11}$$

where $\mathbf{V}_\infty$ is the asymptotic value of $\mathbf{V}$ as $z \to \infty$. Essentially the idea is that (5.10) guarantees the $D(\cdot)$ terms can control the numerator via (5.6), while (5.11) does likewise with the rearrangement

$$< D_{ij}u_iu_j >=< (V_i - V_{i\infty})u_ju_{i,j} >,$$

and use of the arithmetic-geometric mean inequality. What is important is that Galdi & Rionero (1985) show (5.6) *cannot be weakened* by producing a counterexample when $(1 + z)^2$ is replaced by $(1 + z)^{2-\epsilon}$. This, of course, means their existence result for the maximum problem does not hold under weaker behaviour either.

Another possibility to achieve a sensible maximum problem is to have a heat source that ensures a temperature decay like z^{-2} *and* to have a gravity field that decays like z^{-2} at infinity. Such a gravity field is, of course, consistent with Newtonian potential theory. This model is adopted for convection in the infinite region exterior to a sphere by Galdi & Padula (1990).

Whitehead & Chen (1970) report an experimental study together with a linear analysis for a penetrative convection model on a half-space where they achieve a nonlinear basic temperature profile by use of a heat source $Q(z)$. It is interesting to note that all of their basic temperature profiles have the property required above, namely, that the temperature is destabilizing on only a finite layer of the half-space.

This section is concluded by reviewing a piece of work by Dudis & Davis (1971) where again the maximizing solution does not in general exist; it does so for only one value of the coupling parameter, the value they selected. The basic solution whose stability they examine, the buoyancy boundary layer solution of Gill (1966), is an exact solution to the Boussinesq equations of motion. As they point out, Prandtl's "mountain and valley winds in stratified air" contains this solution as a special case when the surface under consideration is vertical rather than slanted.

The fluid occupies the half-space $x > 0$ bounded by an infinite vertical wall at $x = 0$, the acceleration due to gravity acts in the negative $z-$direction (vertically downward) and a stratification is established by a temperature difference ΔT across the half-space, the temperature field varying linearly with z at each z height, thus the boundary conditions are

$$V_i = 0, \qquad T = \Delta T + Gz, \quad \text{at } x = 0,$$
$$V_i \to 0, \qquad T \to Gz, \quad \text{as } x \to \infty, \tag{5.12}$$

where G is a positive constant.

Under the non-dimensionalization of Dudis & Davis (1971) the basic state is

$$U = V = 0, \qquad W = e^{-x}\sin x, \qquad T = e^{-x}\cos x + \frac{2z}{PrRe}. \tag{5.13}$$

The energy equation derived by Dudis & Davis (1971) is:

$$\frac{d}{dt}\left(\frac{1}{2}\|\mathbf{u}\|^2 + \frac{1}{2}\lambda Pr\|\theta\|^2\right)$$
$$= -\left[D(\mathbf{u}) + \lambda D(\theta)\right]$$
$$+ \frac{2}{Re} < \theta w > -Pr\lambda < T_{,i}u_i\theta > -Re < D_{ij}u_iu_j > .$$

They choose $\lambda = 1$ for apparently no good reason, although we point out it does precisely remove the troublesome $< w\theta >$ term, since

$$-\lambda Pr T_{,i}u_i\theta = -\frac{2}{Re}\lambda w\theta - \lambda Pr\frac{\partial T}{\partial x}u\theta.$$

According to theory of Galdi & Rionero (1985) the choice $\lambda = 1$ is necessary and for the very reason that the $< w\theta >$ terms are removed and the energy maximum problem is then well set. (In this case essentially the periodicity is in the y, z directions and T, W decay sufficiently rapidly in x.)

The paper of Dudis & Davis (1971) finds the solution to the Euler-Lagrange equations for the energy maximum problem with $\lambda = 1$, and their results are sharp when compared to linear theory. The paper describes a non-trivial numerical procedure.

5.2 The Salinity Gradient Heated Vertical Sidewall Problem

In a recent piece of work, Kerr (1990) correctly applies energy theory to an interesting half-space problem. He studies the theory of an infinite body of fluid with a vertical salinity gradient heated from a vertical sidewall. In his linear instability analysis, Kerr (1989), the predictions of onset agree well with experimental results. However, it takes a weakly nonlinear analysis, Kerr (1990), to correctly predict that the bifurcation into instability is subcritical and that co-rotating cells are found, in agreement with experiments. He also includes an energy analysis, and it is this aspect that fits into the material of this chapter; and hence we include a description of the relevant work.

The object is to determine nonlinearly a stability criterion for the problem of fluid occupying the half-space $x > 0$, with a vertical linear salinity gradient superimposed and heated at the sidewall $x = 0$. The basic solution of Kerr (1989,1990) is

$$\bar{\mathbf{v}} = (0, 0, \bar{w}(x)), \quad \bar{T} = f(x), \quad \bar{S} = f(x) - z,$$

where $\bar{S}$ is the steady salt concentration, z is in the vertical direction, and $\bar{w}$ and f are determined by (Kerr (1989), p. 329)

$$\bar{w}(x) = \frac{x}{2\sqrt{\pi}} \exp\left(-\frac{1}{4}x^2\right), \quad f'(x) = \frac{1}{\sqrt{\pi}} \exp\left(-\frac{1}{4}x^2\right). \tag{5.14}$$

($\bar{w}$ and f do actually have a time dependence, but this is effectively disregarded in the energy analysis. Also, in the scaling of Kerr (1990) a factor δ^*, which measures the ratio of a vertical to a horizontal length scale, has a presence in (5.14).)

With Pr Prandtl number, τ the salt to thermal diffusivity ratio, and defining a Rayleigh number

$$\mathcal{R} = \frac{g\alpha\Delta T h^3}{\kappa^2}, \tag{5.15}$$

where κ is the thermal diffusivity and h is a vertical length scale, the perturbation equations, in terms of perturbations u_i, p, θ and salt perturbation s, are

$$\frac{\partial u_i}{\partial t} + u_j u_{i,j} + \bar{v}_j u_{i,j} + u_j \bar{v}_{i,j} = -p_{,i} + \mathcal{R}(\theta - s)k_i + Pr\Delta u_i,$$

$$u_{i,i} = 0,$$

$$\frac{\partial\theta}{\partial t} + u_j\theta_{,j} + \bar{v}_j\theta_{,j} + u_j\bar{T}_{,j} = \Delta\theta, \tag{5.16}$$

$$\frac{\partial s}{\partial t} + u_j s_{,j} + \bar{v}_j s_{,j} + u_j \bar{S}_{,j} = \tau\Delta s.$$

These equations are defined on the half-space $\{x > 0\}$ for all positive time. The boundary conditions employed are

$$u_i = \theta = \frac{\partial s}{\partial x} = 0, \quad \text{at } x = 0, \qquad u_i, \theta, s \to 0, \quad \text{as } x \to \infty. \tag{5.17}$$

To derive the energy identity, Kerr (1990) introduces the notation

$$\bar{A}(x,t) = \lim_{K,L \to \infty} \frac{1}{4KL} \int_{-L}^{L} \int_{-K}^{K} A(x,y,z,t)\, dy\, dz,$$

$$<A> = \int_0^\infty \bar{A}(x,t)\, dx.$$

Then, assuming u_i, p, θ, s, and their derivatives as required, belong to the appropriate L^2 class of functions such that the energy equations make sense, he derives four energy identities; namely, those formed by multiplying $(5.16)_1$ by u_i, $(5.16)_3$ by θ, $(5.16)_4$ by s and integrating, and a mixed one formed by multiplying $(5.16)_3$ by s, $(5.16)_4$ by θ and adding. The identities are

$$\frac{1}{2}\frac{d}{dt} <u_i u_i> = - <wu\bar{w}_x> + \mathcal{R}(<\theta w> - <sw>) - Pr <u_{i,j}u_{i,j}>,$$

$$\frac{1}{2}\frac{d}{dt} <\theta^2> = - <\theta u f_x> - <\theta_{,j}\theta_{,j}>,$$

$$\frac{1}{2}\frac{d}{dt} <s^2> = - <su f_x> + <sw> - \tau <s_{,j}s_{,j}>,$$

$$\frac{d}{dt} <s\theta> = - <(\theta + s)u f_x> + <\theta w> - (1+\tau) <s_{,j}\theta_{,j}>.$$

An energy $E(t)$ is defined as

$$E(t) = \frac{1}{2} <u_i u_i> + \frac{1}{2}\lambda_1 <\theta^2> + \frac{1}{2}\lambda_2 <s^2> + \lambda_3 <s\theta>,$$

for coupling parameters $\lambda_1, \lambda_2, \lambda_3\, (> 0)$ to be chosen, and this is observed to satisfy the equation

$$\frac{dE}{dt} = I - \mathcal{D}, \tag{5.18}$$

where

$$\mathcal{D} = Pr <u_{i,j}u_{i,j}> + \lambda_1 <\theta_{,i}\theta_{,i}>$$
$$+ \lambda_2 \tau <s_{,i}s_{,i}> + \lambda_3(1+\tau) <s_{,i}\theta_{,i}>,$$

$$I = - <wu\bar{w}_x> - \lambda_1 <\theta u f_x> - \lambda_2 <su f_x>$$
$$- \lambda_3 <(\theta + s)u f_x> + (\mathcal{R} + \lambda_3) <\theta w> \tag{5.19}$$
$$+ (\lambda_2 - \mathcal{R}) <sw>.$$

The stability criterion is then found to be

$$\sup \frac{I}{D} < 1. \qquad (5.20)$$

Kerr (1990) observes that since the usual Poincaré inequality (with no weight) does not hold for a half-space then the $< \theta w >$ and $< sw >$ terms in (5.19) must not be present, and he selects

$$\lambda_3 = -\mathcal{R}, \qquad \lambda_2 = \mathcal{R} \qquad (5.21)$$

to remove them. Of course, the theory of Galdi & Rionero (1985) requires precisely the choice (5.21) before the maximum problem arising from (5.20) admits a solution; the decay behaviour of $\bar{w}$ and f_x as $x \to \infty$ is very rapid and certainly satisfies the Galdi & Rionero (1985) criteria. Kerr derives a general result from (5.20), but to achieve a stronger result he then assumes the perturbations are periodic in y and z, a restriction that is meaningful since such behaviour is observed experimentally. Kerr (1990, p.543) notes that, ... *in experiments the instabilities are observed to be thin, almost horizontal convection cells with a vertical length scale of order* h, ..., *in none of the experiments was there a layer thickness greater than* h. His further analysis proceeds from (5.18)–(5.20) by splitting u_i, θ, s into two parts, one averaged over the y, z directions, the other being the remainder. By use of a variety of inequalities and estimates, and with the choice $\lambda_1 = (1+\tau^2)/2\tau$, he then arrives at the nonlinear energy stability criterion (Kerr (1990), inequality (3.44)),

$$\frac{1}{2}(1 - \tau)\delta^{*3} + \frac{(1 - \tau)}{\sqrt{Pr\tau}}\mathcal{R}^{1/2}\delta^* < \frac{8\pi^{5/2}}{(\Delta z)^2}, \qquad (5.22)$$

where Δz is the vertical periodicity of the disturbance and δ^* is the ratio of the vertical length scale h to the horizontal length scale $\sqrt{\kappa t}$ (the $\sqrt{t}$ relates to the heating time). Inequality (5.21) is a useful threshold against which the experimental occurrence of subcritical instability may be compared.

6

Generalized Energies and the Lyapunov Method

6.1 The Stabilizing Effect of Rotation in the Bénard Problem

Thus far in convection studies we have explored uses of the energy method that have concentrated on employing some form of kinetic-like energy, involving combinations of L^2 integrals of perturbation quantities. While this is fine and yields strong results for a large class of problem, there are many situations where such an approach leads only to weak results if it works at all, and an alternative device must be sought. In fact, recent work has often employed a variety of integrals rather than just the squares of velocity or temperature perturbations. Drazin & Reid (1981), p. 431, point out that this natural extension of the energy method is essentially the method advocated by Lyapunov for the stability of systems of ordinary differential equations some 60 years or so ago. In this chapter we indicate where a variety of different *generalized energies* (or *Lyapunov functionals*) have been employed to achieve several different effects. Since the applications are usually connected to geophysical or astrophysical problems we discuss this aspect also: chapter 7 is, however, devoted entirely to two geophysical problems where energy theory has proved valuable. One of these, convection in thawing subsea permafrost, is particularly attractive from an energy stability point of view because the construction of a novel *generalized energy* is *necessary* to achieve sharp quantitative stability bounds.

Another good example of a situation where the standard L^2 energy theory yields very conservative results is the rotating Bénard problem, especially for large rotation rates. In the beginning of this chapter we shall review the work of Galdi & Straughan (1985b) who constructed a generalized energy for the Bénard problem in a rotating fluid layer that brings out the stabilizing effect of rotation. Since the analysis differs substantially from that seen so far in the book and the details are somewhat involved, the exposition is necessarily lengthy and divided into §§6.1–6.3.

The physical aspect of convection in a rotating fluid layer is the driving force for analysis. Rossby (1969) reports detailed results for his experiments on the rotating Bénard problem for a wide range of rotation rates (Taylor numbers), for Prandtl numbers, $Pr\, (= \nu/\kappa)$, which correspond to water and to mercury. He observes (p. 329): *Measurements with water revealed two striking features. The first is the presence of a subcritical instability at Taylor numbers greater than 5×10^4 which becomes more pronounced at larger Taylor numbers...The second feature is that water exhibits a maximum heat flux not without rotation but at a Taylor number which is an increasing function of the Rayleigh number...It was conjectured that this increase is due to an 'Ekman-layer-like' modification of the viscous boundary layer.* Clearly, therefore, in the light of Rossby's work, any useful *quantitative* nonlinear analysis is desirable.

The equations for a perturbation (u_i, θ, p) to the stationary solution (referred to a constant rotation reference state) $u_i = 0$, $T = -\beta z + T_0$, are given in Chandrasekhar (1981), p. 87. The fluid is contained in the layer $z \in (0, d)$ and the temperature, T, is kept constant on the bounding planes $z = 0, d$, with $T = T_0$ when $z = 0$, and $T = T_1$ when $z = d$, where $T_0 > T_1$. The quantities u_i, θ, p are the perturbations in velocity, temperature, and pressure (incorporating centrifugal forces and absorbing the constant density), and $\beta = (T_0 - T_1)/d$ is the temperature gradient across the layer.

The energy analysis of Galdi & Straughan (1985b) selects the non-dimensionalization (stars indicating non-dimensional variables):

$$x_i = x_i^* d, \quad u_i = u_i^* U, \quad U = \frac{\nu}{d}, \quad \theta = \theta^* T^\sharp,$$

$$T^\sharp = \left(\frac{Pr\beta}{g\alpha}\right)^{1/2} U, \quad p = p^* P, \quad P = \frac{U\rho_0 \nu}{d}, \quad t^* = t\frac{\nu}{d^2},$$

$$R^2 = \frac{\alpha\beta g d^4}{\kappa\nu}, \quad T = \frac{2d^2\Omega}{\nu}, \quad Pr = \frac{\nu}{\kappa};$$

where the non-dimensional parameters $Ra = R^2$, $\mathcal{T} = T^2$, Pr are the Rayleigh, Taylor, and Prandtl numbers, and Ω is the magnitude of the angular velocity of the layer, which is in the direction $\mathbf{k} = (0, 0, 1)$.

Omitting the stars, the non-dimensional equations for the perturbations are

$$u_{i,t} + u_j u_{i,j} = -p_{,i} + R\theta k_i + \Delta u_i + \epsilon_{ijk} T u_j \delta_{k3},$$

$$u_{i,i} = 0, \tag{6.1}$$

$$Pr(\theta_{,t} + u_i \theta_{,i}) = Rw + \Delta\theta,$$

where the spatial region is the layer $z \in (0, 1)$.

The boundary conditions adopted are that

$$u_i, \theta, p \text{ are periodic in } x, y, \text{ with periods } 2a_1, 2a_2, \text{ respectively.} \tag{6.2}$$

The planes $z = 0, 1$ are *free surfaces* on which no tangential stresses act (cf. Chandrasekhar (1981), pp. 21–22), so that

$$\frac{\partial u}{\partial z} = \frac{\partial v}{\partial z} = w = \theta = 0, \qquad \text{on} \quad z = 0, 1, \tag{6.3}$$

where $\mathbf{u} = (u, v, w)$. To exclude rigid motions we assume the mean values of u, v are zero, i.e.,

$$< u >=< v >= 0,$$

where $< \cdot >$ again denotes integration over a periodicity cell V, and here we take $V = [0, 2a_1) \times [0, 2a_2) \times (0, 1)$. The assumption of free surfaces is not unrealistic. For high rotation rates Ekman layers form and the convecting layer is bounded by the boundary layers that probably act like free surfaces.

In Galdi & Straughan (1985b) a nonlinear energy analysis is developed, which determines a stability boundary $Ra_E = Ra_E(T)$, such that for $R^2 < Ra_E$ we are able to obtain sufficient conditions to guarantee nonlinear stability of the stationary solution given above. This work is described below.

A standard energy method for the rotating Bénard problem employing Joseph's coupling parameter idea, described in chapter 4, would commence with an energy of form

$$E = \frac{1}{2}\|\mathbf{u}\|^2 + \frac{1}{2}\gamma Pr\|\theta\|^2,$$

where $\| \cdot \|$ again denotes the L^2 norm on V and $\gamma (> 0)$ is a coupling parameter to be chosen in such a way as to obtain the sharpest stability boundary. Joseph (1966), pp. 174–175, points out that this method cannot give any information regarding the stabilizing effect of rotation as found by the linear analyses of Chandrasekhar (1953,1981). The reason for this is that the Coriolis term associated with the rotation, the last term in (6.1)$_3$, when multiplied by u_i and integrated over V contributes to the energy equation only via the term

$$T < \epsilon_{ijk}u_i u_j \delta_{3k} > \quad (= 0).$$

In fact, Galdi & Straughan (1985b) show that the optimum result is achieved with $\gamma = 1$, and this is the same as when there is no rotation present. Therefore, the standard energy analysis yields only $Ra_E = 27\pi^4/4$ and no stabilization is observed. When T is of order 10^5 Chandrasekhar (1953,1981) shows that the critical Rayleigh number of linear theory, Ra_L, is of order 21000, and so there is a huge variation between the linear and energy results.

Since the inhibiting effect of rotation on the instability of a fluid layer heated from below has long been recognized as a phenomenon of major importance in Bénard convection and as convection in a rotating system is relevant to many geophysical applications and to such industrial applications as semiconductor crystal growing, it is not surprising that there have been many articles dealing with theoretical or experimental analyses of this problem; see, for example, Chandrasekhar (1953,1981), Chandrasekhar & Elbert (1955), Kloeden & Wells (1983), Langlois (1985), Roberts (1967), Roberts & Stewartson (1974), Rossby (1969), and Veronis (1959,1966,1968b).

Therefore, a nonlinear analysis using an *energy* theory is very desirable. It is clear though that a different energy functional must be chosen to encompass the stabilizing effect of rotation. Vorticity is inherently a quantity associated with rotation and it plays an important role in Chandrasekhar's (1953,1981) linear analysis, and we might expect its inclusion also to be important in any worthwhile energy functional, and this point strongly influenced the choice of *generalized energy* in Galdi & Straughan (1985b) who obtain a nonlinear stability boundary which for Prandtl numbers greater than 1 and Taylor numbers between 10^5 and 10^8 is very close to the experimental results of Rossby (1969) and lies below the curve of linear theory.

To include stabilizing effects in a nonlinear analysis we find it necessary to introduce a generalized energy that involves higher derivatives of u_i, θ and so we need the relevant equations for the higher derivatives. We start with a vorticity component equation and to this end take the curl of $(6.1)_1$ and with $\omega_i = (\text{curl } \mathbf{u})_i$, $\omega = \omega_3$ find

$$\frac{\partial \omega}{\partial t} + u_i \omega_{,i} = \Delta \omega + T w_{,z} + u_{r,y} u_{,r} - u_{r,x} v_{,r}. \tag{6.4}$$

An evolution equation for Δw is obtained by applying the operator curl curl to $(6.1)_1$ and then taking the third component of the resulting equation to find

$$\frac{\partial}{\partial t} \Delta w = R \Delta^* \theta + (u_a u_{i,a})_{,zi} - \Delta(u_i w_{,i}) + \Delta^2 w - T \omega_{,z}, \tag{6.5}$$

where Δ^* denotes $\partial^2/\partial x^2 + \partial^2/\partial y^2$. We also need an equation for $\theta_{,z}$ and this follows by differentiating $(6.1)_3$,

$$Pr \frac{\partial}{\partial t} \theta_{,z} = -Pr(u_{i,z} \theta_{,i} + u_i \theta_{,iz}) + R w_{,z} + \Delta \theta_{,z}. \tag{6.6}$$

In fact, the vorticity only enters the analysis through the combination

$$F = \omega - \frac{T}{R} \xi Pr \theta_{,z},$$

for $0 < \xi < 1$ a coupling parameter to be selected. The evolution equation for F is calculated from (6.4) and (6.6) as

$$\frac{\partial F}{\partial t} = \Delta F + T(1 - \xi) w_{,z} + \frac{\xi T}{R}(Pr - 1)\Delta \theta_{,z}$$
$$+ u_{i,y} u_{,i} - u_{i,x} v_{,i} - u_i F_{,i} + Pr \frac{\xi T}{R} u_{i,z} \theta_{,i}. \tag{6.7}$$

The generalized energy of Galdi & Straughan (1985b) commences with

$$E(t) = \frac{1}{2} \{ \lambda(\|\mathbf{u}\|^2 + Pr\|\theta\|^2) + c_1 Pr\|\theta_{,z}\|^2 + c_2 \|F\|^2 + \|\nabla w\|^2 \},$$

where there are four coupling parameters, $\lambda, c_1, c_2(> 0)$, and $0 < \xi < 1$ to be chosen.

Because equations (6.5)–(6.7) are of higher order than (6.1) extra boundary conditions are needed. The functions $F, \theta_{,z}, \Delta w$ are periodic in x, y with the same periodicities as u_i and additionally, from (6.1)$_3$ and Chandrasekhar (1981), p. 90, eq. (102)$_2$, it follows that

$$F_{,z} = \theta_{,zz} = w_{,zz} = 0, \quad \text{on} \quad z = 0, 1. \tag{6.8}$$

The equation for dE/dt is obtained using (6.1), (6.3), (6.5)–(6.7),

$$
\begin{aligned}
\frac{dE}{dt} = &\lambda\big(2R < \theta w > - < u_{i,j}u_{i,j} > - \|\nabla\theta\|^2\big) + c_1 R < w_{,z}\theta_{,z} > \\
&+ c_2 T(1 - \xi) < w_{,z}F > + T < F_{,z}w > \\
&+ \frac{T^2}{R}\xi Pr < w\theta_{,zz} > + c_1 < \theta_{,z}\Delta\theta_{,z} > \\
&+ c_2 < F\Delta F > + c_2 \frac{T}{R}\xi(Pr - 1) < F\Delta\theta_{,z} > \\
&- R < w\Delta^*\theta > - < w\Delta^2 w > + N,
\end{aligned}
\tag{6.9}
$$

where N contains cubic perturbation quantities and may be arranged to have the form

$$
\begin{aligned}
N = &c_1 Pr < u_i(\theta_{,i}\theta_{,zz} + \theta_{,z}\theta_{,zi}) > - c_2 Pr\xi\frac{T}{R} < F_{,i}u_{i,z}\theta > \\
&+ c_2 < F_{,i}(vu_{i,x} - uu_{i,y}) > + < u_i w_{,i}\Delta w - u_i u_{j,i}w_{,jz} > .
\end{aligned}
\tag{6.10}
$$

The terms on the right of (6.9) are denoted by $I_1 - I_{11}$. We first select the coupling parameters c_1, c_2 to be

$$c_1 = \frac{\xi Pr T^2}{R^2}, \qquad c_2 = \frac{1}{1 - \xi} \ (> 0). \tag{6.11}$$

This choice and use of the boundary conditions (6.2), (6.3) ensures that

$$I_2 + I_5 = 0, \qquad I_3 + I_4 = 0. \tag{6.12}$$

$I_6 - I_{10}$ are now rewritten, after integration by parts and use of (6.2), (6.3), and (6.8), in the form

$$I_6 = -c_1\|\nabla\theta_{,z}\|^2, \qquad I_7 = -c_2\|\nabla F\|^2, \tag{6.13}$$

$$I_8 = -c_2\xi\frac{T}{R}(Pr - 1) < F_{,i}\theta_{,iz} >, \tag{6.14}$$

$$I_9 = R < \theta_{,\alpha}w_{,\alpha} >, \tag{6.15}$$

$$I_{10} = -\|\Delta w\|^2, \tag{6.16}$$

where in (6.15) the subscript α denotes summation over 1 and 2.

Substitution of (6.11)–(6.16) in (6.9) simplifies the partial energy equation to

$$\frac{dE}{dt} = RI - D + N, \tag{6.17}$$

where I and D are the terms

$$I = 2\lambda < \theta w > -c_2\xi\frac{T}{R^2}(Pr - 1) < F_{,i}\theta_{,iz} > + < \theta_{,\alpha}w_{,\alpha} >,$$

$$D = \lambda\{< u_{i,j}u_{i,j} > + < \theta_{,i}\theta_{,i} >\}$$
$$+ c_1 < \theta_{,iz}\theta_{,iz} > + c_2 < F_{,i}F_{,i} > +\|\Delta w\|^2.$$

We wish to develop an energy analysis along the same lines as that described in chapter 4. In the case of (6.17) the cubic nature of the non-linearities necessitates the introduction of a stronger dissipation term D to control the nonlinear terms in N.

6.2 Construction of a Generalized Energy

The cubic nonlinearities in N contain terms like $\theta_{,zz}$ and $w_{,iz}$, and we control these by introducing a dissipative term that contains higher order derivatives, such as $\|\Delta\mathbf{u}\|^2$ and $\|\Delta\theta\|^2$. We, therefore, define another part, $E_1(t)$, of the energy by

$$E_1(t) = \frac{1}{2} < u_{i,j}u_{i,j} > +\frac{1}{2}Pr < \theta_{,i}\theta_{,i} > . \tag{6.18}$$

This part of the energy plays no role in the variational problem for determining the critical Rayleigh number, Ra_E, of energy theory, it serves simply to dominate the nonlinear terms.

We differentiate E_1 and use the boundary conditions to find

$$\frac{dE_1}{dt} = < u_{i,j}u_{i,jt} > +Pr < \theta_{,i}\theta_{,it} >,$$
$$= - < u_{i,t}\Delta u_i > -Pr < \theta_{,t}\Delta\theta > . \tag{6.19}$$

The right-hand side of (6.19) is manipulated by multiplying $(6.1)_1$ by Δu_i and $(6.1)_3$ by $\Delta\theta$ and after several integrations by parts and further use of the boundary conditions we obtain

$$\frac{dE_1}{dt} = F_1 - D_1 + N_1,$$

where F_1, D_1, N_1 are given by

$$F_1 = 2R < w_{,i}\theta_{,i} >, \qquad D_1 = \|\Delta\theta\|^2 + < \Delta u_i \Delta u_i >,$$
$$N_1 = < u_j u_{i,j} \Delta u_i > + Pr < u_i \theta_{,i} \Delta\theta > .$$

The final energy functional employed by Galdi & Straughan (1985b) has the form

$$\mathcal{E} = E(t) + bE_1(t),$$

where $b\,(> 0)$ is yet another coupling parameter to be selected judiciously. It is important to reiterate that the two parts to $\mathcal{E}(t)$ play different roles. The part $E(t)$ is the important one that yields the stability boundary and its choice is crucial, whereas the part $E_1(t)$ plays the role of a piece to dominate nonlinearities. In applying a generalized energy technique such as this to other problems the difficult part is always to select the $E(t)$ piece in such a way that a sharp stability boundary is achieved. There may be several ways to choose the E_1 component to dominate the nonlinear terms. For example, L^p integrals with p other than 2, are employed in chapter 7 on the thawing subsea permafrost problem, and may also be used in convection studies in electrohydrodynamics and in ferrohydrodynamics, see chapters 11 and 12.

The energy equation for $\mathcal{E}(t)$ may be written

$$\frac{d\mathcal{E}}{dt} = RI - D + bF_1 - bD_1 + N + bN_1. \tag{6.20}$$

The energy maximum problem involves only those quadratic terms arising from E, i.e., I and D, and so we define

$$\frac{1}{R_E} = \max_{\mathcal{H}} \frac{I}{D}, \tag{6.21}$$

where the space of admissible functions is

$$\mathcal{H} = \{F, u_i, \theta | u_{i,i} = 0, \ u_i, \theta, F \text{ periodic in } x, y \text{ and satisfy (6.3), (6.8)}\}.$$

The following inequality then results directly from the energy equation (6.20):

$$\frac{d\mathcal{E}}{dt} \leq -D\left(1 - \frac{R}{R_E}\right) + bF_1 - bD_1 + N + bN_1. \tag{6.22}$$

The stability threshold is R_E and so we require $R < R_E$ and we set $m = R/R_E$, where $0 < m < 1$.

To proceed from (6.22) we must estimate F_1 and the N, N_1 terms. A bound for F_1 is achieved as follows. An integration by parts shows,

$$< u_{i,j} u_{i,j} > = - < u_i \Delta u_i >,$$
$$\leq \|\mathbf{u}\|\,\|\Delta \mathbf{u}\|,$$

where in the last line the Cauchy-Schwarz inequality has been used. The Wirtinger inequality (see Appendix 1)

$$\pi\|\mathbf{u}\| \leq \|\nabla\mathbf{u}\|,$$

then allows us to deduce

$$\pi\|\nabla\mathbf{u}\| \leq \|\Delta\mathbf{u}\|.$$

With this inequality F_1 may be estimated as

$$bF_1 \leq \frac{1}{2}bD_1 + 2b\frac{R^2}{\pi^2}\|\nabla\theta\|^2.$$

A similar estimation of the $\|\nabla\theta\|$ term allows us to further show

$$bF_1 \leq \frac{1}{2}bD_1 + 2b\frac{R^2}{\pi^4 c_1}D. \tag{6.23}$$

We now choose b to eliminate the F_1 term and here select

$$b = \frac{\pi^4 c_1(1-m)}{4R^2},$$

and then (6.23) in (6.22) yields

$$\frac{d\mathcal{E}}{dt} \leq -\mathcal{D} + bN_1 + N,$$

where $\mathcal{D}$ has been defined by

$$\mathcal{D} = \frac{1}{2}bD_1 + \frac{1}{2}(1-m)D.$$

It is important to note that because we are assuming $R < R_E$, $m < 1$, and so the coefficient of D is positive.

The estimation of the N and N_1 terms proceeds via taking out the supremum of one function and then employing the Cauchy-Schwarz inequality, e.g.,

$$bN_1 \leq b \sup_V |\mathbf{u}| \left(\|\nabla\mathbf{u}\|\,\|\Delta\mathbf{u}\| + Pr\|\nabla\theta\|\,\|\Delta\theta\|\right).$$

We further know (Appendix 1),

$$\sup_V |\mathbf{u}| \leq C\|\Delta\mathbf{u}\|.$$

Therefore, for example, employing this estimate in the inequality for N_1, we derive

$$bN_1 \leq C\sqrt{\frac{2}{b}}(1+Pr)bD_1\mathcal{E}^{1/2},$$

$$\leq 2C\sqrt{\frac{2}{b}}(1+Pr)\mathcal{D}\mathcal{E}^{1/2}.$$

The estimate for N is similar and proceeds again taking out the supremum of a function and estimating it in terms of the L^2 norm of the Laplacian. By estimating N and N_1 in this way, in terms of $\mathcal{D}\mathcal{E}^{1/2}$, we derive from the energy inequality,

$$\frac{d\mathcal{E}}{dt} \leq -\mathcal{D}(1 - A\mathcal{E}^{1/2}), \tag{6.24}$$

where A is a constant that behaves like $b^{-1/2}$.

Inequality (6.24) yields sufficient conditions to ensure that the generalized energy *decays monotonically to zero* (c.f. chapter 2, §3). To see this, suppose that $\mathcal{E}(0) < A^{-2}$. Then, from (6.24) $d\mathcal{E}/dt < 0$ in some neighbourhood of 0, and so $\mathcal{E}$ cannot exceed its initial value for all $t > 0$; hence,

$$\frac{d\mathcal{E}}{dt} \leq -B\mathcal{D}, \tag{6.25}$$

where $B = 1 - A\mathcal{E}^{1/2}(0)$. Due to the boundary conditions on u_i, θ and the fact that

$$< F >= 0,$$

by use of the Wirtinger inequality (Appendix 1),

$$\mathcal{D} \geq (1 - m)\pi^2 Pr^{-1}\mathcal{E},$$

and after using this in (6.25) and integrating we may show that

$$\mathcal{E}(t) \leq \mathcal{E}(0) \exp\left\{-\pi^2[1 - A\mathcal{E}^{1/2}(0)](1 - R/R_E)t/Pr\right\}, \tag{6.26}$$

from which it follows that $\mathcal{E}(t) \to 0$ as $t \to \infty$ in a rapid, monotonic manner.

The importance in arranging A to behave at least like $b^{-1/2}$ is to ensure the size of the initial data need not become vanishingly small as $\mathcal{T}$ increases. The case $\mathcal{T}$ small is not explicitly dealt with here but may be handled by a slightly different analysis, see Galdi & Straughan (1985b).

6.3 Determination of the Nonlinear Stability Threshold

To determine the nonlinear stability boundary, Ra_E, it remains to resolve the maximum problem (6.21). By taking the variation in (6.21), the Euler-Lagrange equations for the maximum are found to be

$$R_E \delta_{i3}(2\lambda - \Delta^*)\theta + 2\lambda\Delta u_i - 2\Delta^2 w\delta_{i3} = -p_{,i}, \tag{6.27}$$

$$c_2\xi\frac{T}{R^2}R_E(Pr - 1)\Delta\frac{\partial}{\partial z}\theta + 2c_2\Delta F = 0, \tag{6.28}$$

$$R_E(2\lambda - \Delta^*)w - c_2\xi\frac{T}{R^2}R_E(Pr-1)\Delta\frac{\partial}{\partial z}F + 2\left(\lambda - c_1\frac{\partial^2}{\partial z^2}\right)\Delta\theta = 0, \quad (6.29)$$

where p is a Lagrange multiplier introduced because of the solenoidal constraint on $\mathbf{u}$.

To solve the above system for the lowest eigenvalue R_E we reduce to a single equation in w to find

$$2T\frac{\xi}{R^2}\left[\frac{1}{2}\xi R_E^2\frac{(Pr-1)^2}{(1-\xi)R^2} - 2Pr\right]\Delta^3(\lambda - \Delta^*)\frac{\partial^2}{\partial z^2}w$$
$$+ 4\lambda\Delta^3(\lambda - \Delta^*)w - R_E^2\Delta^*(2\lambda - \Delta^*)^2w = 0. \quad (6.30)$$

This equation is linear and we seek a solution of the form

$$w = e^{i(mx+ny)}W(z). \quad (6.31)$$

This does not restrict us to a rectangular plan form for the cell since a simple change in form transforms to other cell shapes, see Drazin & Reid (1981), pp. 54–56, and chapter 3.

Because we are dealing with two free surfaces, the boundary conditions that W must satisfy are

$$W = D^{(2n)}W = 0, \quad \text{on} \quad z = 0,1, \quad n = 1, 2, \ldots,$$

where $D = d/dz$. Therefore, in (6.31), we need only consider

$$W(z) = A\sin r\pi z, \quad r = 1, 2, \ldots.$$

Put (6.31) with the above form for W into (6.30) and solve for R_E^2 to obtain, with $a^2 = m^2 + n^2$ and $\Lambda_1 = (\lambda + a^2)(r^2\pi^2 + a^2)^3$,

$$R_E^{-2} = \frac{a^2(2\lambda + a^2)^2(1-\xi)R^4 + \xi^2T(Pr-1)^2\Lambda_1r^2\pi^2}{4R^2\Lambda_1(\xi PrTr^2\pi^2 + \lambda R^2)(1-\xi)}. \quad (6.32)$$

From (6.32) it may be shown that to find the stability boundary we must solve the following quadratic in R_E^2 :

$$R_E^4 - R_E^2\, 4\lambda\frac{(\lambda + a^2)}{(2\lambda + a^2)^2}\frac{(\pi^2r^2 + a^2)^3}{a^2}$$
$$= T\frac{r^2\pi^2(\lambda + a^2)}{(2\lambda + a^2)^2}\frac{(\pi^2r^2 + a^2)^3}{a^2}f(\xi), \quad (6.33)$$

where we have set

$$f(\xi) = 2\xi\left[2Pr - \frac{\xi(Pr-1)^2}{2(1-\xi)}\right].$$

In equation (6.33),

$$R_E^2 = R_E^2(a^2, r^2; \lambda, \xi; Pr, \mathcal{T}),$$

where a^2, r^2 are parameters arising from the variational treatment, λ, ξ are energy (coupling) parameters, and $Pr, \mathcal{T}$ are given.

It is of interest to note that with no rotation, i.e., $\mathcal{T} = 0$, (6.33) reduces to the classical situation

$$R_E^2 = \frac{(\pi^2 r^2 + a^2)^3}{a^2},$$

for $\lambda \to \infty$.

In the general case, we need the largest value of R_E^2 in terms of λ, ξ, whereas for fixed λ, ξ we minimize R_E^2 in a^2, r^2. First, choose ξ to maximize R_E^2. This requires $f(\xi) = 4$ if $Pr \geq 1$ and $f(\xi) = 4Pr^2$ when $Pr < 1$. With this choice of $f(\xi)$ in (6.33) it is not difficult to show R_E^2 is minimized by the selection $r = 1$. Whence, it is sufficient to restrict attention to those R_E^2 that satisfy

$$R_E^2 = \lambda g(\lambda, a^2) + \sqrt{\lambda^2 g^2(\lambda, a^2) + 2\mathcal{T}\pi^2 g(\lambda, a^2)}, \qquad (6.34)$$

where we have set

$$g(\lambda, a^2) = \frac{2(\lambda + a^2)(\pi^2 + a^2)^3}{(2\lambda + a^2)^2 a^2},$$

and where we have considered only $Pr \geq 1$; if $Pr < 1$ the factor $\mathcal{T}\pi^2$ is replaced by $\mathcal{T}\pi^2 Pr^2$ in (6.34).

For $\mathcal{T}$ fixed it remains to determine

$$Ra_E = \max_\lambda \min_{a^2} R_E^2.$$

Before describing the general results for Ra_E we digress to consider an asymptotic feature, namely, from (6.34), setting $\Lambda = \pi^2 + a^2$, for $0 < \lambda << 1$,

$$R_E^2 \sim \frac{2\mathcal{T}\pi\Lambda^{3/2}}{a^2} + \lambda\left[\frac{2\Lambda^3}{a^4} - \frac{3\mathcal{T}\pi\Lambda^{3/2}}{a^4}\right] + O(\lambda^2). \qquad (6.35)$$

We require R_E^2 as large as possible (in λ) and so for λ very small the coefficient of λ in (6.35) should be non-negative. The minimum of the $O(1)$

term is when $a^2 = 2\pi^2$ and so for the coefficient of λ in (6.35) to be positive we have

$$\mathcal{T} < 12\pi^4 \approx 1168.9.$$

For $\mathcal{T}$ larger than this value we pick $\lambda = 0$, $a^2 = 2\pi^2$, to find

$$R_E^2 = 3^{3/2}\pi^2 \mathcal{T} \approx 51.28\mathcal{T},$$

for $\mathcal{T} \geq 1168.9$.

The numerical results for Ra_E, for $Pr \geq 1$, are indicated in Figures 6.2 and 6.3. A comparison of the nonlinear energy stability results with those of linear theory, Chandrasekhar (1981), those of Veronis (1968b), and with the experimental reults of Rossby (1969) is revealing. Figure 6.1 shows the linear instability curve, Chandrasekhar (1981), against the experimental curve of Rossby (1969); while Rossby's experiments were with fixed boundaries, he argues that rotation allows Ekman layers to form, and these then serve to relax the rigid boundary conditions so the fluid "sees" free boundaries.

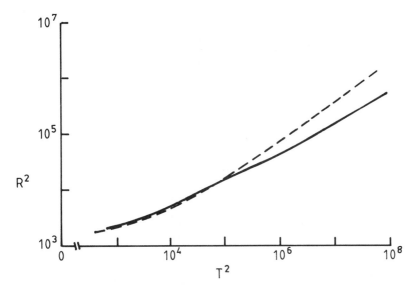

Figure 6.1. Critical Rayleigh number against Taylor number for two fixed surfaces. The broken line indicates the linear results; solid line represents Rossby's results. (After Galdi & Straughan (1985b).)

Figures 6.2 and 6.3 compare the linear results, Chandrasekhar (1981), which are the same as those found by Veronis (1968b), with the energy results, for two free surfaces.

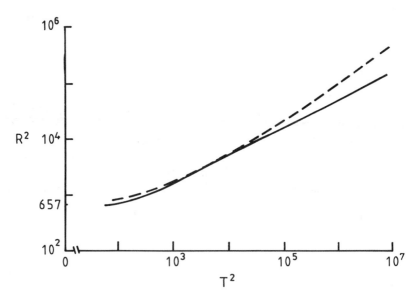

Figure 6.2. Critical Rayleigh number against Taylor number for two free surfaces. The broken line indicates the linear results; solid line represents energy results. (After Galdi & Straughan (1985b).)

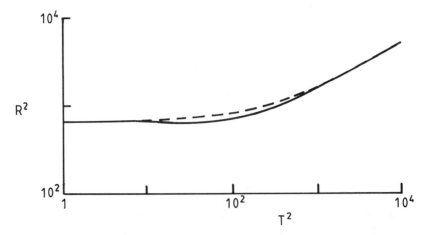

Figure 6.3. Critical Rayleigh number against Taylor number for two free surfaces, small Taylor number. The broken line indicates the linear results; solid line represents energy results. (After Galdi & Straughan (1985b).)

It is very interesting to observe that the qualitative behaviour of the energy bound in Figure 6.2 is very similar to that found in Rossby's (1969) experiments. Of course, the energy results do not prove that sub-critical instabilities occur, they simply allow us to predict that for a Rayleigh number less than Ra_E we have conditional energy stability against *three-dimensional* disturbances, in a precise sense.

Mulone & Rionero (1989) continue the analysis of §§6.2 and 6.3. They adopt the same functional as Galdi & Straughan (1985b) but choose different coupling parameters to explore other ranges of Taylor numbers. The generalized energy method described in §6.2 has been extended to the Bénard problem with a superimposed shear flow by Mulone (1990). This is an interesting paper, and he necessarily has to devise a different functional from that of §6.2, although the gist of the technique is that of this chapter.

6.4 Penetrative and Non-Boussinesq Convection

A pioneering piece of work on penetrative convection is the beautiful paper of Veronis (1963). Our description of penetrative convection is largely based on his paper.

In many natural phenomena the process of thermal convection involves a penetrative motion into a stably stratified fluid. Examples of penetrative convection are to be found in several areas of geo- and astrophysical fluid dynamics. For example, the Earth's atmosphere is bounded below by the ground or ocean; this bounding surface is heated by solar radiation, and the air close to the surface then becomes warmer than the upper air, and so a gravitationally unstable system results. When convection occurs, the warm air rises and penetrates into regions that are stably stratified.

In the oceans evaporation is the main cause of gravitational instability near the sea surface. As the cool surface water is carried downward, it too enters regions that are stably stratified. In a star the surface layer is stable. According to Veronis (1963) p. 641, *at some distance below the "surface" the increase in temperature due to adiabatic compression causes negative hydrogen to form and the latter is opaque to photons. The temperature gradient therefore rises to a value greater than the adiabatic gradient and the region is unstable. Depending on the type of star, this superadiabatic gradient can extend far into the interior to a point where the very high temperature causes the gas to become completely ionized and the gradient no longer is superadiabatic. An unstable layer is formed, therefore, with stable fluid both above and below.*

To study penetrative convection on a laboratory scale Veronis (1963) proposed a simplified model. Consider a layer of water with the bottom maintained at 0°C and the top at a temperature greater than 4°C. Due to the fact that the density of water below 4°C is a decreasing function of temperature, the above situation results in a gravitationally unstable layer of fluid lying below a stably stratified one. When convection occurs in the lower layer the motions will penetrate into the upper layer.

The phenomenon of penetrative convection is lucidly described by Veronis (1963) who analysed in detail the linearized system and developed a weakly nonlinear finite amplitude analysis for two stress free boundaries. Veronis (1963) finds that penetrative convection differs markedly from the classical case of Bénard convection for which sub-critical instabilities are not possible and shows that a finite amplitude solution exists for sub-critical Rayleigh numbers; in particular (p. 662), he writes, *The results indicate that when the bottom of a layer of water is maintained at 0°C temperature and the top > 4°C, a finite amplitude instability should set in at values of the Rayleigh number below the critical value given by linear stability theory.*

The bifurcation analysis of Veronis (1963) is based on a power series method and as Busse (1978) observes, such expansions do not represent a suitable approach to derive quantitative information about convection beyond the immediate neighbourhood of the critical Rayleigh number; however, Busse also argues that power series methods are still the only way to analyse the variety of *three-dimensional* solutions to the perturbation equations.

The phenomenon of penetrative convection and, in particular, applications to geophysics and to astrophysics is still a subject of active research, see e.g., Azouni (1983), Azouni & Normand (1983a,b), George et al. (1989), Kondo & Unno (1982, 1983), Payne et al. (1988), Walden & Ahlers (1981), Whitehead (1971), Whitehead & Chen (1970), and Zahn, Toomre & Latour (1982).

The equations of motion in the layer differ from those derived by the Boussinesq approximation in §3.2, but only in the form of the dependence of density on temperature in the body force term. The density is still assumed constant everywhere except in the body force term. Because the density of water is almost parabolic in the range $0 - 8°C$ Veronis (1963) chooses as equation of state,

$$\rho = \rho_0 \left[1 - \alpha(T - 4)^2 \right], \tag{6.36}$$

where $\rho(T)$ is the density, ρ_0 the density at 4°C, $\alpha \approx 7.68 \times 10^{-6} \, (°C^{-2})$, and Veronis points out that even at 14°C (6.36) involves only a 10% error.

Suppose the fluid is contained in the layer $z \in (0, d)$ and the surface temperatures are

$$T = 0°C \quad \text{at} \quad z = 0, \qquad T = T_2 \quad \text{at} \quad z = d,$$

where T_2 is a constant temperature not less than 4°C, and the phase change effect at the lower boundary is ignored. The perturbation equations to the motionless solution

$$\bar{u}_i = 0, \quad \bar{T} = \beta z, \quad \beta = \frac{T_2}{d},$$

are in non-dimensional form

$$u_{i,t} + u_j u_{i,j} = -p_{,i} + \Delta u_i - 2R\theta(\xi - z)k_i + Pr\theta^2 k_i, \quad (6.37)$$
$$u_{i,i} = 0, \quad (6.38)$$
$$Pr(\theta_{,t} + u_i \theta_{,i}) = -Rw + \Delta\theta, \quad (6.39)$$

where R^2 is a Rayleigh number, $\xi = 4/T_2$, and $k_i = \delta_{i3}$.

The energy analysis of Straughan (1985) proceeds via an L^2 method using a Sobolev inequality to handle the resulting cubic nonlinear term; however, this analysis is only conditional, c.f. §2.3. In §2.4 it was shown for the one-dimensional equation with a quadratic force that boundedness could be achieved by employing a weighted energy. Payne & Straughan (1987), in fact, employ a weighted energy precisely to overcome the conditional restriction. They select a *generalized energy* of the form

$$E(t) = \frac{1}{2} < u_i u_i > + \frac{1}{2} Pr < (\mu - 2z)\theta^2 >,$$

in which $\mu \, (\geq 2)$ is a coupling parameter to be chosen optimally. The next step is to differentiate E and substitute for $u_{i,t}, \theta_{,t}$ from (6.37), (6.39). In the standard energy technique the $\theta u_i \theta_{,i}$ term that arises integrates to zero, but the effect of the weight in this case gives rise to a term

$$-Pr < \theta^2 w >,$$

which exactly cancels the corresponding destabilizing term in the momentum equation. Having removed the offending cubic term, the *weighted* energy equation then has the same form as usual,

$$\frac{dE}{dt} = -\mathcal{D} + IR,$$

where now,

$$\mathcal{D} = < u_{i,j} u_{i,j} > + < (\mu - 2z)\theta_{,i}\theta_{,i} >$$

and

$$I = - < (\mu + 2\xi - 4z)\theta w > .$$

The energy limit R_E is defined in the usual way,

$$\frac{1}{R_E} = \max_{\mathcal{H}} \frac{I}{\mathcal{D}},$$

and again it may be shown that $R < R_E$ is a sufficient condition for non-linear stability. We stress that the stability obtained is *unconditional;* i.e., for all initial amplitude disturbances, this being achieved by the use of the spatial weight in the definition of $E(t)$. The unconditional nonlinear energy stability boundary is, in fact, very close to the linear instability one; correct numerical results are given in Payne & Straughan (1988).

In general, nonlinear equations of state of the form (6.36) form part of the theory of so-called non-Boussinesq convection. Non-Boussinesq convection essentially employs the Boussinesq approximation in the sense that the only variations in density are through the body force, but the equation of state is no longer linear in the temperature.

In a recent series of developments, several studies have questioned whether the quadratic equation (6.36) is accurate enough for detailed comparison with field studies and experiments involving convection in a layer where the fluid has a density maximum, see e.g., Merker et al. (1979) and the references therein. These writers suggest using a density law like

$$\rho = \rho_0[1 + AT - BT^2 + CT^3], \tag{6.40}$$

or even like

$$\rho = \rho_0[1 + AT - BT^2 + CT^3 - DT^4 + ET^5], \tag{6.41}$$

where ρ_0 now denotes the density of water at $0°C$ and the coefficients A, B, C, D, E are constants obtained by curve fitting to data points. Merker et al. (1979) suggest for water, values in (6.40) of

$$A = 6.85650 \times 10^{-5} \ (°C^{-1}),$$
$$B = 8.82063 \times 10^{-6} \ (°C^{-2}),$$
$$C = 4.16668 \times 10^{-8} \ (°C^{-3}),$$

while in (6.41)

$$A = 6.79939 \times 10^{-5} \ (°C^{-1}),$$
$$B = 9.10749 \times 10^{-6} \ (°C^{-2}),$$
$$C = 1.00543 \times 10^{-7} \ (°C^{-3}),$$
$$D = 1.12689 \times 10^{-9} \ (°C^{-4}),$$
$$E = 6.59285 \times 10^{-12} \ (°C^{-5}).$$

Merker et al. (1979) further regard (6.41) as *exact* in the range $0°C$ to $40°C$. It is important to note that for highly accurate predictions on the onset of convection Merker et al. (1979) suggest (*on the basis of linear theory*) that (6.40) is about 10% more accurate than (6.36) whereas (6.41) yields approximately a 3% improvement over (6.40). The numerical results from the weighted energy analysis of McKay & Straughan (1990) suggest that it is preferable to employ (6.40), but (6.41) for such a small gain in accuracy leads to much greater mathematical complications. We take this opportunity to point out that (6.40) has been employed by Niedrauer & Martin (1979) in their investigation of the convective motion of brine in channels formed in sea ice. Equation (6.40) has also been advocated by Ruddick & Shirtcliffe (1979).

Merker et al. (1979) give only linear instability results. However, the analytical methods of Veronis (1963) clearly indicate that sub-critical in-stabilities will exist for (6.40) and (6.41); Busse (1967) points this out in the general case where either the thermal expansion coefficient, or the vis-cosity depend on temperature. Thus a nonlinear energy stability result is desirable in that it represents a useful *quantitative threshold* (i.e., lower bound) for the hexagon sub-critical instability curve drawn qualitatively by Busse (1967), p. 644. The above objective of estimating the turning point of "inverted bifurcation" has been investigated *experimentally* in the recent work of Walden & Ahlers (1981) in liquid Helium I and of Azouni & Normand (1983a) in water.

We now describe analysis appropriate to (6.40); further details, and anal-ysis pertinent to (6.41) may be found in McKay & Straughan (1990). Sup-pose the boundary conditions for the temperature are

$$T = T_1 \quad \text{at} \quad z = 0, \qquad T = T_2 \quad (\geq 4°C) \quad \text{at} \quad z = d,$$

where T_1 and T_2 are constants with $0 \leq T_1 \leq 4$. The equations for thermally driven convective motion in a layer are (3.39), and we use them with $\rho(T)$ in the buoyancy term given by (6.40). The resulting system then possesses the motionless solution,

$$\bar{u}_i = 0, \qquad \bar{T} = T_1 + \beta z, \qquad \beta = \frac{\Delta T}{d}, \qquad \Delta T = T_2 - T_1, \qquad (6.42)$$

with the hydrostatic pressure determined from what amounts to $(3.39)_1$ but with the buoyancy density law (6.40).

Under a suitable non-dimensionalization the equations for a perturbation to this solution are

$$u_{i,t} + u_j u_{i,j} = -p_{,i} + \Delta u_i - R f_1 \theta k_i$$
$$+ Pr k_i \left(f_2 \theta^2 - a_2 \frac{Pr}{R} \theta^3 \right), \qquad (6.43)$$
$$u_{i,i} = 0, \qquad (6.44)$$
$$Pr(\theta_{,t} + u_i \theta_{,i}) = -Rw + \Delta \theta, \qquad (6.45)$$

where $Ra = R^2$ is the Rayleigh number, $\xi = T_1 / \Delta T$, and

$$f_1(z) = 1 - 2a_1(\xi + z) + 3a_2(\xi + z)^2, \qquad f_2(z) = a_1 - 3a_2(\xi + z),$$

with a_1, a_2 being defined by

$$a_1 = \frac{B}{A}\Delta T, \qquad a_2 = \frac{C}{A}(\Delta T)^2.$$

The standard boundary conditions apply, i.e., $u_i = \theta = 0$ on $z = 0, 1$, and u_i, θ, p are periodic in x, y.

It is quite possible that exchange of stabilities holds for the linearized version of (6.43)–(6.45). Indeed, Merker et al. (1979) *assume* $\sigma \in \mathbf{R}$. For two free surfaces it is easy to adapt the Spiegel method to prove exchange of stabilities, c.f. §4.1.

We may obtain L^2 energy identities from (6.43) and (6.45), and these are

$$\frac{1}{2}\frac{d}{dt}\|\mathbf{u}\|^2 = - R < f_1\theta w > - < u_{i,j}u_{i,j} >$$
$$+ Pr < f_2\theta^2 w > - a_2\frac{Pr^2}{R} < w\theta^3 >,$$

$$\frac{1}{2}Pr\frac{d}{dt}\|\theta\|^2 = - R < w\theta > - < \theta_{,i}\theta_{,i} > .$$

I do not see how to proceed *directly* from here since I do not see a way of dominating the $< w\theta^3 >$ term (arising from the cubic term in the density) by means of the stabilizing terms $< u_{i,j}u_{i,j} >$ and $< \theta_{,i}\theta_{,i} >$. Instead, two approaches using a generalized energy are suggested.

One alternative is to employ a weighted energy, the other defines an energy that also includes a piece of the L^4 integral of θ to control the relevant destabilizing term. The addition of a piece of the L^4 integral is somewhat analogous to the approach used to construct the generalized energy of §6.2. The L^4 part is used only to dominate the nonlinear terms, while the stability boundary is determined by that part of the energy involving the L^2 integrals. (It is worth noting that (6.41) leads to worse problems and evidently a yet more complicated energy must be employed which involves either both L^4 and L^6 integrals of θ, or a more involved weight function.)

For (6.43)–(6.45) we may choose,

$$E(t) = \frac{1}{2}\|\mathbf{u}\|^2 + \frac{1}{2}\lambda Pr\|\theta\|^2 + \frac{\mu}{4}Pr\|\phi\|^2,$$

where $\phi = \theta^2$, and λ, μ are positive coupling parameters, then conditional nonlinear stability can be achieved. To obtain *unconditional* nonlinear stability one approach is to utilize the weighted energy

$$E(t) = \frac{1}{2}\|\mathbf{u}\|^2 + \frac{1}{2}Pr\mu_1\|\theta\|^2 + \frac{1}{3}Pr < \mu_2(z)\theta^3 > + \frac{\lambda}{4R^2}Pr\|\phi\|^2,$$

where

$$\mu_2(z) = \frac{3}{PrR}(\lambda + a_2 Pr^2)z.$$

The coupling parameters μ_1 and λ have to be opportunely selected. The numerical results so far available are very satisfactory, at least in the range $0 - 9°C$.

6.5 Thermohaline Convection with Heat and Salt Below

We return to equations (4.72) and for convenience suppose ϕ represents a perturbation to a salt concentration field, C, and we take $\xi = 0$. In equations (4.72) with $\xi = 0$ if the layer is salty above and heated below then both effects are destabilizing, $H = \tilde{H} = -1$, $\hat{H} = 1$ in (4.72) and the linearized system is symmetric, therefore, the linear and nonlinear boundaries coincide and no sub-critical instabilities can occur. This result was first established by Shir & Joseph (1968). If, however, the layer is salted below, which is a stabilizing effect, while the layer is simultaneously heated from below, which is a destabilizing effect, then the two opposing effects make an energy analysis decidedly more complicated, to achieve sharp results. For this situation the equations are (4.72) but with $H = -1$, $\tilde{H} = \hat{H} = +1$, and $\xi = 0$. For precisely this problem Joseph (1970) introduced a new twist into the theory of energy stability; the idea of a *generalized energy*. He sets $\tau = Pc/Pr$, $R_s = \alpha R$ and then chooses

$$E(t) = \frac{1}{2}\|\mathbf{u}\|^2 + \frac{1}{2}\frac{Pr}{1+\tau}\left[\|\gamma\|^2 + \|\psi\|^2\right],$$

where

$$\gamma = \lambda_1\theta - \lambda_2\phi, \qquad \psi = \lambda_1\theta - \tau\lambda_2\phi,$$

and where λ_1, λ_2 are coupling parameters linked by the equation

$$\lambda_1 + \frac{1}{\lambda_1} - \frac{2\alpha\lambda_2}{(1+\tau)} = \alpha\left(\frac{1}{\lambda_2} - \lambda_2\right) + \frac{2\lambda_1\tau}{1+\tau}.$$

The above choice of energy is evidently necessary to produce a sharp result. Indeed, Joseph (1970) finds a stability boundary that is very close to the linear instability one. He indicates that sub-critical instabilities arise in precisely the region delimited by his energy analysis.

From the point of view of Lyapunov functionals, or generalized energies, in fluid mechanics, the paper of Joseph (1970) is a very important one. It is the first one I know of where such a *generalized energy* is employed *very effectively* to achieve sharp results evidently not attainable by a standard energy analysis.

Another noteworthy energy analysis for the thermohaline problem is that of Kerr (1990), already mentioned in chapter 5: this is further complicated by the fact that the problem is there posed on a half-space.

6.6 Bio-Convection and Use of a Spatial Weight to Symmetrize

We now describe an energy analysis for a suspension of swimming micro-organisms, partly because it is of interest for its own sake, but also because the choice of generalized energy is suggested directly by previous work *on the linear problem*. This illustrates an important point: if there are any special features present in the linear problem, it may be possible to exploit them to suggest an opportune generalized energy.

The continuum model of bio-convection we describe is due to Childress et al. (1975) and Levandowsky et al. (1975), while the energy analysis was given by Galdi & Straughan (1985a).

A brief explanation of the physical picture is as follows. Suppose that a suspension of micro-organisms is contained in a fluid layer, say the infinite layer between the planes $z = -H, 0$, and suppose that a gravitational field acts in the negative z direction. The organisms have a density greater than that of the containing fluid and also have a natural tendency to swim in the upward (increasing z) direction. If a sufficient number of organisms are present, eventually the situation arises where the upper layer of the fluid is dominated by micro-organisms. These in turn, being of a density greater than the fluid, will tend to fall under the action of gravity. Hence an instability somewhat akin to Rayleigh-Taylor instability may develop. The striking thing about this instability is that it does not happen in a haphazard manner; rather, the organisms tend to fall in discrete "chimneys" in an ordered pattern, although several pattern types are possible, see Childress et al. (1975), Levandowsky et al. (1975). Another interesting convection mechanism that exhibits a distinct chimney structure, albeit with rising rather than falling plumes, is that caused by one component of a mixture being frozen out of a solution; a continuum theory for this phenomenon has been derived by Loper & Roberts (1978,1980). Details relevant to the Earth's core are provided by Loper & Roberts (1981), and related material may be found in Hills et al. (1983), Hills & Roberts (1987a,1987b).

The basic model we employ for the motion of micro-organisms is derived by Childress et al. (1975). Let $c(\mathbf{x}, t)$ denote the concentration of micro-organisms in the suspension. The conservation law for the organisms is easily written in terms of a flux, $\mathbf{J}$, given by

$$J_i = cU(c, z)\delta_{i3} - D_{ij}c_{,j} \,, \tag{6.46}$$

where U is the upward swimming velocity of the organisms and $\mathbf{D}$ is the diffusion tensor given by

$$\mathbf{D} = \begin{pmatrix} \kappa_1(c, z) & 0 & 0 \\ 0 & \kappa_1(c, z) & 0 \\ 0 & 0 & \kappa(c, z) \end{pmatrix}, \tag{6.47}$$

in which κ and κ_1 are positive functions of the indicated arguments.

The governing equations, based on an incompressible linear viscous fluid model, are in the continuum approximation,

$$
\begin{gathered}
u_{i,t} + u_j u_{i,j} = -\frac{1}{\rho} p_{,i} - g(1 + \alpha c)\delta_{i3} + \nu \Delta u_i, \\
u_{i,i} = 0, \\
c_{,t} + u_i c_{,i} = -J_{i,i},
\end{gathered}
\tag{6.48}
$$

where ρ is the constant density of the suspension, u_i the velocity, p the pressure, ν the viscosity, g is gravity, and α is a positive constant that expresses the ratio of density of a micro-organism to that of the growth medium.

The linear stability of two classes of equilibrium solution is considered in Childress et al. (1975), namely:

Case I.
$$
U = U_0, \quad \kappa = \kappa_0, \quad \kappa_1 = \delta \kappa_0,
$$

where U_0, κ, δ are constant.

Case II.

κ/U not explicitly dependent on z, κ_1 arbitrary.

Galdi & Straughan (1985a) analyse a subclass of the more general Class II. They need the boundary conditions:

$$
J_i n_i = 0, \quad u_i = 0 \quad \text{when} \quad z = -H, 0.
$$

The former expresses the condition that no material flows out of the bounding surfaces $z = -H, 0$, whereas the latter is the no-slip condition. To ensure uniqueness of the equilibrium solution it is also necessary to impose the following restriction on the mean concentration:

$$
<c> \overset{\text{def}}{=} c_n = \text{const.}
$$

The subclass of solutions of Case II investigated involves those for which κ, U depend only on z, $dU/dz \leq 0$, $U/\kappa = h^{-1}$ (constant), and κ_1 is arbitrary, apart from being continuously differentiable. The basic equilibrium solution of this subclass (denoted by an overbar) is

$$
\bar{u}_i = 0, \quad \bar{c} = K(z) = c_0 \exp\left(\frac{U_0 z}{\kappa_0}\right), \quad z \in [-H, 0],
$$

with $c_0 = K(0)$, and U_0, κ_0 the values of U, κ at $z = 0$.

It is important to note that the basic equilibrium solution is *nonlinear* in z, so that an energy stability analysis is likely to be different from that required for such constant gradient problems as the standard Bénard one.

To study stability we set

$$\mathbf{u} = (u, v, w) = (u_1, u_2, u_3), \qquad c = K(z) + \phi(\mathbf{x}, t), \qquad p = \bar{p}(z) + \pi,$$

where (u_i, ϕ, π) are perturbations to the equilibrium values $(\bar{u}_i, K, \bar{p})$. The key to the choice of energy arises from the fact that Childress et al. (1975) observed that the growth rate of linear theory could be shown to be real by essentially using K' as a weight. In the light of this, we derive the non-dimensionalized perturbation equations to (6.48) but first divide the perturbation equation arising from (6.48)$_3$ by K' to obtain

$$u_{i,t} + \sigma^{-1} u_j u_{i,j} = -\pi_{,i} + \Delta u_i - R\phi\delta_{i3},$$

$$u_{i,i} = 0,$$

$$\frac{\sigma\phi_{,t} + u_i\phi_{,i}}{K'} = -Rw - \frac{(\phi U)'}{K'} + \frac{(\kappa_1\phi_x)_x + (\kappa_1\phi_y)_y}{K'} + \frac{(\kappa\phi')'}{K'},$$

(6.49)

where $R^2 = (\alpha g c_0 h^3 / \nu \kappa_0)$ is like a Rayleigh number, $\sigma = \nu/\kappa_0$ is a constant Schmidt number, $h = \kappa_0/U_0$ is a constant unit of length, $(0, \lambda)$ is now the non-dimensional layer, and $'$ denotes $\partial/\partial z$.

The boundary conditions are

$$\kappa\phi' = \phi U, \quad u_i = 0 \qquad \text{when} \qquad z = 0, \lambda, \tag{6.50}$$

together with

$$u_i, \phi, p \text{ are periodic functions in } x, y. \tag{6.51}$$

The disturbance cell is denoted by V.

In the form (6.49), the linear operator L that acts on (u, v, w, ϕ) is given by

$$\begin{pmatrix} \Delta & 0 & 0 & 0 \\ 0 & \Delta & 0 & 0 \\ 0 & 0 & \Delta & -R \\ 0 & 0 & -R & \frac{1}{K'}\left[\partial_\alpha(\kappa_1\partial_\alpha\cdot) + (\partial_z(\kappa\partial_z\cdot) - \partial_z(U\cdot)\right] \end{pmatrix}, \tag{6.52}$$

where the repeated α signifies summation over $\alpha = 1, 2$. With the aid of (6.50) and (6.51) it is easily verified that L is symmetric, thanks to the weight K'.

The natural energy to use is suggested by (6.49), namely,

$$E(t) = \frac{1}{2} < u_i u_i > + \frac{1}{2}\sigma\left\langle \frac{\phi^2}{K'} \right\rangle.$$

Differentiating E and substituting from (6.49), with use of the boundary conditions (6.50), (6.51), we find the energy equation is

$$\frac{dE}{dt} = -2R < \phi w > - < u_{i,j}u_{i,j} > - \left\langle \frac{\kappa_1}{K'}(\phi_x^2 + \phi_y^2) \right\rangle$$

$$- \left\langle \frac{\kappa}{K'}\left(\phi_z - \frac{\phi}{h}\right)^2 \right\rangle - \frac{1}{2h}\left\langle \frac{w\phi^2}{K'} \right\rangle. \tag{6.53}$$

Since the linear part of the system is symmetrized we may appeal to the result of chapter 4, following Lemma 4.1, which shows the linear and non-linear stability boundaries are the same for a symmetric system, provided we can handle the cubic nonlinear term in (6.53). Due to the presence of the last term in (6.53), the one in $w\phi^2$, the energy decay result derived by Galdi & Straughan (1985a) is achieved through use of a Sobolev inequality and is only conditional.

6.7   Convection in a Porous Vertical Slab

We complete this chapter by looking at another problem where a standard energy may not reveal as much information as a generalized one, and, in particular, where a *natural* generalized energy is suggested by previous work on the linearized problem.

Kassoy (1980) describes several situations, together with relevant references, of geophysical problems where the mathematical description is provided by studying convection in a vertical porous slot. In an interesting paper Gill (1969) showed that when a temperature difference is imposed across the slot, no convection is possible in the sense that the basic solution is *linearly* stable to two-dimensional disturbances no matter how large the Rayleigh number. In Straughan (1988) Gill's (1969) linear analysis is used to motivate the choice of a generalized energy, which shows that three-dimensional nonlinear disturbances will always decay provided the initial amplitude is less than a threshold, which behaves like the inverse of the Rayleigh number.

The model uses the Boussinesq approximation and Darcy's law, see Gill (1969). The velocity $\mathbf{v} = (u, v, w)$ is referred to Cartesian axes x_i chosen so that the z-axis points vertically upward and such that the boundaries at temperature $T = \pm\frac{1}{2}$ are given by $x = \pm\frac{1}{2}$. The steady solution to the problem is

$$u = v = 0, \quad w = x, \quad T = x.$$

The equations for a perturbation to this solution are

$$u_i = -p_{,i} + \delta_{i3}\theta, \qquad u_{i,i} = 0,$$
$$Ra(\theta_{,t} + u_i\theta_{,i} + u + x\theta_{,3}) = \Delta\theta, \tag{6.54}$$

where Ra is the Rayleigh number. The boundary conditions are:

$$\theta = u = 0, \qquad x = \pm\frac{1}{2}; \qquad u_i, \theta, p \text{ periodic in } y, z.$$

The development of a generalized energy analysis commences by multiplying $(6.54)_3$ by $\Delta\theta$ and integrating over V to obtain

$$\frac{1}{2}Ra\frac{d}{dt}\|\nabla\theta\|^2 = -\|\Delta\theta\|^2 + Ra < u\Delta\theta >$$
$$+ Ra < x\theta_{,3}\Delta\theta > + Ra < u_i\theta_{,i}\Delta\theta > . \tag{6.55}$$

Next, take curlcurl of $(6.54)_1$ to see that

$$-\Delta u_i = \delta_{j3}\theta_{,ij} - \delta_{i3}\Delta\theta; \qquad (6.56)$$

the first component of this is

$$-\Delta u = \theta_{,13}. \qquad (6.57)$$

Integrating by parts and using the boundary conditions, we may transform the third term on the right of (6.55) to

$$
\begin{aligned}
< x\theta_{,3}\Delta\theta > &= - < x\theta\Delta\theta_{,3} > \\
&= \frac{1}{2} < (x\theta_{,i}\theta_{,i})_{,3} > + < \theta x_{,i}\theta_{,i3} > \\
&= < \theta\theta_{,13} > \\
&= - < \theta\Delta u > \\
&= - < u\Delta\theta >,
\end{aligned}
$$

substituting from (6.57) in the second last step. This relation is used to reduce (6.55) to

$$\frac{1}{2}Ra\frac{d}{dt}\|\nabla\theta\|^2 = -\|\Delta\theta\|^2 + Ra < u_i\theta_{,i}\Delta\theta > . \qquad (6.58)$$

Gill's result follows easily from (6.58) since in the linear theory the last term is not present.

To deal with the nonlinearity we write

$$< u_i\theta_{,i}\Delta\theta > \le \sup_V |\mathbf{u}|\, \|\nabla\theta\|\, \|\Delta\theta\|. \qquad (6.59)$$

If we now also suppose $u_i n_i = 0$ on the lateral cell boundary (a condition that automatically holds for a finite region), then we have the inequality used in §6.2, see Appendix 1,

$$\sup_V |\mathbf{u}| \le C\|\Delta\mathbf{u}\|.$$

We use this inequality in (6.59) and put the result in (6.58) to find

$$\frac{1}{2}Ra\frac{d}{dt}\|\nabla\theta\|^2 \le -\|\Delta\theta\|^2 + RaC\|\Delta\mathbf{u}\|\, \|\nabla\theta\|\, \|\Delta\theta\|. \qquad (6.60)$$

We now need a bound for $\|\Delta u\|$, and to this end we square (6.56) to derive

$$\|\Delta\mathbf{u}\|^2 = < \Delta\theta(\theta_{,11} + \theta_{,22}) > \le \|\Delta\theta\|^2, \qquad (6.61)$$

where we have used the boundary conditions and the fact that $\theta_{,3} = 0$ on $x = \pm\frac{1}{2}$. Inequality (6.61) applied to (6.60) leads to

$$\frac{1}{2}Ra\frac{d}{dt}\|\nabla\theta\|^2 \leq \|\Delta\theta\|^2(RaC\|\nabla\theta\| - 1). \qquad (6.62)$$

We now require

$$\|\nabla\theta(\mathbf{x},0)\| < \frac{1}{RaC}, \qquad (6.63)$$

and then (6.62) together with the inequality, see Appendix 1, $\|\Delta\theta\| \geq \pi\|\nabla\theta\|$, allows us to deduce that $\|\nabla\theta(\mathbf{x},t)\|^2 \to 0$ at least exponentially as $t \to \infty$.

Furthermore, Poincaré's inequality allows us to deduce the same decay for $\|\theta\|^2$. Also, since from (6.54)$_1$

$$\|\mathbf{u}\|^2 = <\theta w> \leq \frac{1}{2}\Big(\|\theta\|^2 + \|w\|^2\Big),$$

we see that

$$\|\mathbf{u}\|^2 \leq \|\theta\|^2,$$

and hence $\|\mathbf{u}\|^2$ must also decay at least exponentially.

To sum up, we have demonstrated that provided the initial data satisfy the restriction (6.63), there is always nonlinear stability. Since Ra is known and C is computable, (6.63) represents a useful practical bound. However, if the initial temperature gradients exceed the bound in (6.63) then one cannot rule out the possibility of a finite amplitude instability.

7
Geophysical Problems

7.1 Patterned (or Polygonal) Ground Formation

In this chapter we describe two geophysical problems where energy theory has proved very useful. Not only has an application of energy theory yielded useful information, but also, the mathematics of the problem has *necessitated* the introduction of novel generalized energies.

We begin with patterned ground formation, a subject developed analytically in the first instance by Ray et al. (1983), Gleason (1984), and Gleason et al. (1986). Early theoretical studies identified some of the processes involved, e.g., Nordenskjold (1909), Low (1925), Gripp (1926), and Washburn (1973,1980). A theory that involves a five step process was proposed by Professor R.D. Gunn in George et al. (1989), and we describe this work; partly, because it would appear the process description is the most complete so far, but also, because this is where nonlinear energy stability theory was first employed in the subject.

Polygonal ground, which in this description consists of stone borders forming regular hexagons with soil centres, represents one of the most striking and interesting small scale geological phenomena. In fact, the regularity in size and shape of polygonal ground is often remarkable. However, the existence of these features is not widely known because they generally exist above the timberline in the high mountains of the temperature zone or in remote locations far beyond the treeline in the Arctic and Antarctic. The stone borders usually form regular hexagons that are all about the same size at a single location. Between different sites, polygonal stone nets vary from about 10cm in width to more than 4m, as may be seen from Table 7.1 of actual width to depth measurements.

In Gleason's (1984) thesis, pp. 117–120, he includes three tables of data. Data appropriate to hexagonal cells on land, Tables 7.1 and 7.2 are reproduced here. His second table concerns field study data for patterned ground that has formed *under water* in shallow lakes in Mono Craters,

California, and in Medicine Bow, Wyoming. Finally, he includes data for sorted stripes. We include only that for hexagonal patterned ground formed on land since this is relevant to our analysis. The other data is, however, important in its own right and is proving valuable in theoretical treatment, by G. McKay and the writer, of patterned ground formation under water.

Site Location	USGS Quadrangle	Width	Depth	Elevation	I
Niwot Ridge	Ward, CO	2.30	0.53	3500	a
Niwot Ridge	Ward, CO	3.80	1.00	3500	a
Caribou Mt.	Ward, CO	1.70	0.48	3650	a
Caribou Mt.	Ward, CO	1.70	0.48	3650	a
Arikaree Glacier	Monarch Lake, CO	0.86	0.18	3800	a
Arikaree Glacier	Monarch Lake, CO	0.83	0.16	3800	a
Albion Saddle	Ward, CO	0.40	0.10	3650	a
Albion Saddle	Ward, CO	0.50	0.13	3650	a
Albion Saddle	Ward, CO	3.90	1.10	3650	a
Medicine Bow Peak	Medicine Bow, WY	2.30	0.70	3650	a
Medicine Bow Peak	Medicine Bow, WY	2.90	0.70	3650	a
Trail Ridge Road	Trail Ridge	4.00	1.10	3660	a
Green Lake No. 4	Ward, CO	2.30	0.71	3550	a
Chief Mountain	Franks Peak, WY	2.30	0.83	3400	b
Beartooth Plateau	Beartooth Butte & Alpine, MT	0.20	0.07	3400	b
Dana Plateau	Mono Craters, CA	0.64	0.16	3500	b
Parker Pass Creek	Mono Craters, CA	0.70	0.14	3400	b
Mt. Hare Region Richardson Mountains, Yukon Terr.	Eagle River, Canada series A 502 map 1161	0.16	0.05	1150	c

Table 7.1. Field study data for polygonal
ground (hexagons) in the USA. Width,
depth, and elevation are in metres.
Column I denotes Investigator: a=Ray; b=Krantz & Gunn;
c=Gleason & Gunn. (After Gleason (1984).)

Site Location	Country	Width	Depth	Elevation	I
Macquarie Is.	Australia	0.25	0.80	300	a
Hafravatn	SW Iceland	1.46	0.37	110	b
Hraunhreppur	W Iceland	0.48	0.12	30	c
Latraheidi	NW Iceland	2.63	0.69	180	c
Thorskafjardarheidi	NW Iceland	2.03	0.50	450	c

Table 7.2. Field study data for polygonal
ground (hexagons) in Australia and Iceland.
Width, depth, and elevation are in metres.
Column I denotes Investigator: a=Caine;
b=Stingl; c=Schunke. (After Gleason (1984).)

Presently, theory requires three a priori conditions before stone polygons
may form. The first of these is the existence of alternating freeze-thaw
cycles within the soil. For the hexagonal patterns studied here these are
annual cycles that follow the rhythm of the seasons throughout the year.
The second condition necessitates that the soil must be saturated with
water for at least part of the year. The final requirement is that an impermeable frozen soil barrier must underlie the active layer, that is, the layer
of 'soil that alternately freezes and thaws. For the annual cycles considered here, permafrost must be present to form the impermeable frozen-soil
barrier. When the three conditions outlined above are satisfied, the formation of polygonal ground follows a five step process, described completely
by George et al. (1989). Professor R.D. Gunn has, in fact, succeeded in
growing stone polygons in the laboratory by reproducing these five steps.
The five stages are now described.

*Step 1. Permeability Enhancement as a Result of the Formation of Needle
Ice and Frost Heaving in the Soil.* For the onset of convective motion of
water in the soil, Step 2, a critical permeability must be present in the soil
before polygonal ground can form. Usually, only sand or gravel possess
permeabilities sufficiently high. Silty soils, such as those where patterned
ground has been found, produce a large amount of frost heaving, but have
low permeabilities. The necessary increase in permeability is produced by
the formation of needle ice and ice lenses. Needle ice, Washburn (1973), is
an accumulation of slender, bristle-like ice crystals found in soils subjected
to freeze-thaw cycles. The elongation of these ice crystals is perpendicular
to the permafrost surface, and the needles are commonly a centimetre or
more in length. Since the needles consist of essentially pure ice, their growth

thrusts aside soil and rocks. After a few freeze-thaw cycles, the soil volume may increase by more than a hundred percent, and the entire surface of the ground is thrust upward to accommodate this expansion, Embleton & King (1975). When the soil thaws, the melting of the ice crystals leaves the ground with a greatly enhanced porosity, and this in turn leads to greatly increased permeability. A silty soil thus tends to increase in permeability with each freeze-thaw cycle until the critical permeability is reached, then Step 2 of the patterned ground process commences.

During field studies by Professors Krantz and Gunn and their co-workers, they observed that soil from the central part of the stone polygons is quite silty and would not have a sufficiently large permeability to sustain patterned ground formation. Any attempt to transport these soils to the laboratory collapses the porous structure left by the needle ice and ice lenses. Laboratory measurements then confirm permeabilities well below the value required for the formation of polygonal ground. The same soils, when subjected to multiple freeze-thaw cycles in laboratory experiments, have been found to expand and frost heave upwards until after about thirty cycles they begin to form polygonal ground.

Step 2. Onset of Buoyancy Driven Natural Convection in the Water Saturated Soil. The active layer is that layer of soil above the permafrost that thaws each summer and freezes solid during the winter. In the thawed summer state, water in the active layer near the permafrost interface remains close to its freezing point, $0°C$; the water in the soil nearer the surface is relatively much warmer. As we saw in chapter 6, water is unusual in that it has a maximum density at $4°C$. Thus, in the active layer, warmer, denser water near $4°C$ overlies colder, less dense water near its freezing point. When the permeability is high enough and the permafrost is sufficiently melted that the active layer is deep enough, heavier water sinks with consequently lighter water rising to set up convection currents in the soil. The convection currents are responsible for fixing the size and shape of the stone polygons, which arise through Steps 3 and 4.

Step 3. Formation of a Pattern in the Permafrost Interface. The convection currents in the active layer set up a pattern of hexagonal convection cells. The water rises up the centre of the hexagon and flows down along the cell boundary. Downflow carries relatively warmer water from near the surface toward the impermeable permafrost interface, and this induces melting along the sides of the hexagons. In the centre of a cell colder water is carried toward the surface and this slows down melting. This process results in a series of isolated frozen soil peaks in the centre of the hexagon surrounded by an interconnected (hexagonal) continuous trough in the permafrost surface.

Step 4. Formation of Polygonal Ground Through Frost Heaving. In regions
where there is vigorous frost, the process of frost heaving will push any
rocks originally in the soil slowly to the surface. A well known example
of frost heaving is the appearance of stones in a recently ploughed field.
The type of soils in which a frost heave effect is predominant tend to be
of a silty character. Corte's (1966) experiments showed that in a rock-
soil mixture subjected to freeze-thawing the larger stones will be pushed
upwards leaving finer material below. Rocks do not necessarily move in the
vertical direction, but rather, they move at ninety degrees to the freezing
front, see e.g., Washburn (1973). In active patterned ground formations,
stones are displaced perpendicularly to the frozen soil interface, and so
move upward and sideways away from the peaks at the centre of a cell
and congregate near the hexagonal cell boundary that moves more slowly
upward. This aspect is the actual reason for the hexagonal stone pattern
since the stones are following the border of the hexagonal cell formed by
downflow at the permafrost interface.

One argument is that frost heaving of stones is not produced by a stationary frozen surface nor can the freezing come from above since water is
present above the permafrost until the time when the soil is frozen rigid.
In order that the freezing action arise from below it is necessary for the
permafrost to be substantially colder than the freezing point of water and
then heat will be conducted away from the ice-water interface which will
then freeze in an upward direction. The permafrost temperature at 5–10m
below the surface is approximately equal to the average annual temperature
of the air at the surface and so this leads to the conclusion that hexagonal
patterned ground formations are not likely to be found in abundance at
places with average annual temperatures greater than about $-5°C$. These
conclusions were also drawn by Goldthwaite (1976).

Field observations and laboratory experiments by the groups associated
with Professors Gunn and Krantz evidently verify the theory above. For example, some permafrost is present in the mountains of Wyoming at heights
around 3000m, where the average annual air temperature is close to $0°C$.
But, only a few faint and inactive stone patterns are found at these elevations and these are probably relics of the last ice age. In contrast, a
considerable amount of patterned ground has been discovered at elevations
of about 3600m, see Table 7.1, and here the average annual air temperature
is approximately $5°C$ colder. In Professor Gunn's laboratory experiments,
stone polygons did not form even after 500 freeze-thaw cycles when the
temperature of the simulated permafrost was held close to $0°C$. Stone
polygon formation was, however, observed after only 20–30 freeze-thaw cycles when the lower portion of the simulated permafrost was maintained at
a temperature of $-5°C$.

Step 5. Perpetuation of the Hexagonal Pattern. Rocks have a higher thermal conductivity than soil, and so once stones have begun to concentrate

over the troughs in the permafrost, these troughs become self-perpetuating due to accelerated melting caused by the higher thermal conductivity of rock. Eventually, the convective motion of water in the active layer may cease because the removal of the stones from the central part of the hexagonal cell causes the permeability to decrease. There still remains a tendency for further segregation of stone because of the continuing presence of an undulating permafrost surface that is now perpetuated by the higher thermal conductivity of rocks concentrated along the cell borders.

It is believed that the above five-step process, proposed by Professor R.D. Gunn, produces polygonal stone nets that are characterized by regularity in their size and shape. Previous work on the subject of patterned ground is thoroughly reviewed by Ray et al. (1983) and George et al. (1989).

7.2 A Mathematical Model for Patterned Ground Formation

We suppose the porous material (the active layer) occupies the infinite layer $z \in (0, d)$, with the temperature of the lower plane maintained at $0°C$ whereas the temperature of the upper plane is kept fixed at a temperature above $4°C$. Due to the fact that the porous material contains water whose density below $4°C$ is a decreasing function of temperature, the situation envisaged results in a gravitationally unstable layer lying below a stably stratified one. When convection occurs in the lower layer the motions will penetrate into the upper layer, as discussed in the context of a fluid in §6.2.

The equations we employ utilize Darcy's law, although for the density in the body force term we employ the Veronis form, (6.36), namely,

$$\rho = \rho_0\{1 - \alpha(T - 4)^2\}, \tag{7.1}$$

where ρ_0 is the density at $4°C$ and $\alpha \approx 7.68 \times 10^{-6}$ ($°C^{-2}$). The basic equations for the fluid motion in the active layer are then (4.5)–(4.7), with $Q \equiv 0$ and with the body force term in (4.5) replaced by (7.1).

The boundary conditions we select are

$$k_i v_i = 0 \quad \text{at} \quad z = 0, d,$$
$$T = 0°C \quad \text{at} \quad z = 0, \tag{7.2}$$
$$\delta_1 k_i T_{,i} + \delta_2 T = c \quad \text{at} \quad z = d.$$

The number c is a prescribed constant, and δ_1, δ_2 are constants given in terms of a *radiation parameter* a by

$$\delta_1 = \frac{1}{1+a}, \qquad \delta_2 = \frac{a}{1+a}.$$

The boundary condition $(7.2)_3$ is so written, since it allows us to write the (steady) conduction solution in terms of the temperature of the upper surface in the steady state and this quantity features in the non-dimensionalization. Condition $(7.2)_3$ is only approximate since a movement of the boundary is occuring due to the phase change; although we believe it is acceptable to neglect this movement since the timescale for convection is much shorter. Also, condition $(7.2)_3$ allows us to examine the important limiting cases of prescribed heat flux, $a = 0$, and prescribed constant temperature, $a \to \infty$.

The (steady) conduction solution, which then results to (4.5)–(4.7), $(Q \equiv 0)$, (7.1) and (7.2) is

$$v_i = 0, \qquad \bar{T} = \beta z, \tag{7.3}$$

where $\beta = T_1/d$ and the upper surface temperature T_1 is related to c by

$$T_1 = \frac{cd(1+a)}{1+ad}.$$

The steady pressure field is

$$\bar{p}(z) = p_0 - \rho_0 g z + \frac{\alpha \rho_0 g}{3\beta} (\beta z - 4)^3, \tag{7.4}$$

where p_0 is some conveniently chosen pressure reference scale.

To reflect the increased permeability due to needle ice formation, George et al. (1989), chose a linear permeability relation of form

$$k(z) = k_0(1 + \gamma z),$$

and define

$$f(z) = \frac{k_0}{k} = \frac{1}{1 + \gamma z}.$$

Under a non-dimensionalization into the layer $z \in (0,1)$, defining the Prandtl number $Pr = d^2 \mu / k_0 \rho_0 \kappa$, a parameter $\xi = 4/T_1$, and

$$R = \sqrt{\frac{g\alpha\rho_0\beta^2 d^3 k_0}{\kappa\mu}},$$

the non-dimensionalized perturbation equations, for a perturbation to the steady solution (7.3), (7.4), are

$$Au_{i,t} = -p_{,i} - fu_i - 2R\theta(\xi - z)k_i + Pr\theta^2 k_i,$$
$$u_{i,i} = 0, \tag{7.5}$$
$$Pr(\theta_{,t} + u_i\theta_{,i}) = -Rw + \Delta\theta,$$

where $w = u_3$. George et al. (1989) define a Rayleigh number, Ra, to reflect the depth of the layer, which in the conducting state is actually destabilizing, by

$$Ra = \xi^3 R^2. \tag{7.6}$$

The boundary conditions (7.2) for the perturbation quantities become

$$
\begin{aligned}
w &= 0 \quad \text{at} \quad z = 0, 1, \\
\theta &= 0 \quad \text{at} \quad z = 0, \\
\frac{\partial \theta}{\partial z} + a\theta &= 0 \quad \text{at} \quad z = 1;
\end{aligned}
\tag{7.7}
$$

and we also assume $\mathbf{u}, \theta, p$ have a *periodic* structure in (x, y), such as one consistent with hexagonal convection cells.

Linear Instability. The Spiegel method of establishing exchange of stabilities, discussed in §4.1, applies to the linearized version of (7.5), (7.7), provided $f = $ constant. Thus, due to this fact, and since the energy results are very close to the linear ones, George et al. (1989) examine only stationary convection, which reduces to finding the smallest eigenvalue R, of the system

$$
\begin{aligned}
p_{,i} &= -fu_i - 2R\theta(\xi - z)k_i, \\
u_{i,i} &= 0, \\
Rw &= \Delta\theta.
\end{aligned}
\tag{7.8}
$$

The usual normal mode method with

$$\theta = \Theta(z)e^{i(mx+ny)}, \qquad w = W(z)e^{i(mx+ny)},$$

with the wave number k given by $k^2 = m^2 + n^2$ and $D = d/dz$, reduces (7.8) to

$$
\begin{aligned}
(D^2 - k^2)\Theta &= RW, \\
(D^2 - k^2)W &= -\frac{f'}{f}DW + \frac{2k^2 R}{f}(\xi - z)\Theta,
\end{aligned}
\tag{7.9}
$$

where $-f'/f = \gamma/(1 + \gamma z)$. This system was solved numerically by the compound matrix method, see Appendix 2, and the minimum of $R^2(k^2)$ was found by golden section search. The relevant boundary conditions are:

$$
\begin{aligned}
W &= 0 \quad \text{at} \quad z = 0, 1, \\
\Theta &= 0 \quad \text{at} \quad z = 0, \\
D\Theta + a\Theta &= 0 \quad \text{at} \quad z = 1.
\end{aligned}
\tag{7.10}
$$

As emphasized thoroughout the book, linear theory gives a Rayleigh number boundary, which if exceeded ensures instability. It does not preclude the possibility of subcritical instabilities. Energy theory was applied to penetrative convection in a porous medium by George et al. (1989) who find results that are very close to those obtained by linear theory. This shows that linear theory has essentially captured the physics of the onset of pore water convection, the process we believe responsible for determining the aspect ratio of the patterned ground cells.

Nonlinear Energy Stability. To investigate the nonlinear stability of the steady solution (7.3), (7.4) we use the energy

$$E = \frac{1}{2}A\|\mathbf{u}\|^2 + \frac{1}{2}\lambda Pr\|\theta\|^2, \tag{7.11}$$

where $\lambda\,(>0)$ is a coupling parameter, $\|\cdot\|$ is the $L^2(V)$ norm, with V a disturbance cell.

We differentiate E, use (7.5) and the boundary conditions to derive the energy equation

$$\begin{aligned}
\frac{dE}{dt} =& - <f|\mathbf{u}|^2> -\lambda D(\theta) \\
& - a\int_\Gamma \theta^2 dA - R<\theta w(2\xi + \lambda - 2z)> +Pr<\theta^2 w>,
\end{aligned} \tag{7.12}$$

where $D(\cdot)$ is the Dirichlet integral, $<\cdot>$ represents the integral over V, and Γ is that part of the boundary of V that lies in $z = 1$.

Define now

$$\mathcal{D} = \lambda D(\theta) + <f|\mathbf{u}|^2> + a\int_\Gamma \theta^2 dA, \tag{7.13}$$

$$I = - <\theta w(2\xi + \lambda - 2z)>, \tag{7.14}$$

and from (7.12) we obtain

$$\frac{dE}{dt} \le -\mathcal{D}R\left(\frac{1}{R} - \frac{1}{R_E}\right) + Pr<\theta^2 w>, \tag{7.15}$$

where

$$\frac{1}{R_E(\lambda)} = \max_{\mathcal{H}} \frac{I}{\mathcal{D}}, \tag{7.16}$$

$\mathcal{H}$ being the space of admissible functions.

Nonlinear conditional stability follows from inequality (7.15) in a manner similar to that described in §2.3. Hence, suppose $R < R_E$, define $b = R^{-1} - R_E^{-1}\,(>0)$, and then with the aid of the Cauchy-Schwarz inequality we derive from (7.15),

$$\frac{dE}{dt} \le -bR\mathcal{D} + Pr\|\theta^2\|\,\|w\|. \tag{7.17}$$

The Sobolev inequality
$$\|\theta^2\| \le c_1 D(\theta),$$
is now used in (7.17) to show

$$\frac{dE}{dt} \le -bR\mathcal{D}\left(1 - \frac{c_1 Pr\sqrt{2}}{bR\lambda\sqrt{A}}\sqrt{E}\right). \qquad (7.18)$$

Hence, if

$$\text{(i)}\quad R < R_E, \qquad \text{(ii)}\quad E^{1/2}(0) < \frac{bR\lambda\sqrt{A}}{c_1 Pr\sqrt{2}}, \qquad (7.19)$$

then a calculation similar to that in §2.3 shows that $E(t) \to 0$ as $t \to \infty$ at least exponentially.

We must now determine R_E. The Euler-Lagrange equations for the maximum in (7.16) are found after setting $\phi = \lambda^{1/2}\theta$ to remove the λ dependence from the $D(\theta)$ term in $\mathcal{D}$,

$$\begin{aligned} 2\Delta\phi - R_E M(z)w &= 0, \\ 2fu_i + R_E M(z)\phi k_i &= p_{,i}, \end{aligned} \qquad (7.20)$$

where div $\mathbf{u} = 0$, p is a Lagrange multiplier, and

$$M(z) = \frac{2\xi + \lambda - 2z}{\sqrt{\lambda}}. \qquad (7.21)$$

Equations (7.20) are to be solved in conjunction with the boundary conditions (7.7).

The system (7.20), (7.7) is again reduced by normal modes to an eigenvalue problem for a system of ordinary differential equations and

$$Ra_E = \xi^3 \max_{\lambda} \min_{k^2} R_E^2(\lambda, k^2) \qquad (7.22)$$

is found numerically by using the compound matrix method, the optimization routine using golden section search.

For the energy stability analysis of patterned ground formation the conditional bound (7.19)(ii) is not a strong restriction. In terms of the velocity perturbation we require

$$\|\mathbf{u}(0)\| < \frac{bR\bar{\lambda}}{c_1 Pr},$$

where $\bar{\lambda}$ denotes the best value of λ in (7.22). It is important to observe that this bound is independent of A, which means that the velocity condition is not as restrictive as the temperature one. Physically this is good, since we expect the fluid velocity to drive the cellular motion.

7.3 Results and Conclusions for Patterned Ground Formation

Extensive tabulated values of critical Rayleigh numbers of linear theory, Ra_L, and of energy theory, Ra_E, for various a values and changing permeability (varying γ) are given in George et al. (1989). Figures 7.1 to 7.3 show the variation of Ra with the upper surface temperature for various permeabilities, variation of Ra with a ($\gamma = 0$), and the variation of the wave number with a ($\gamma = 0$).

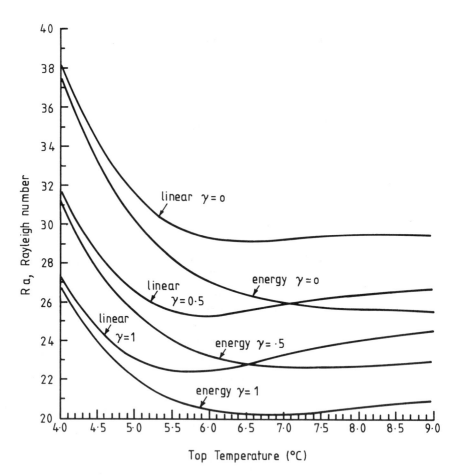

Figure 7.1. Ra against top temperature, $\gamma = 0, 0.5, 1$, with the energy curve being the lower curve of each pair. (After George et al. Copyright 1989 by Gordon and Breach. Reprinted with permission.)

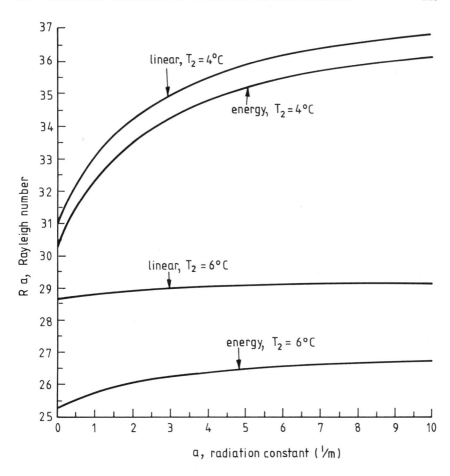

Figure 7.2. Ra against a, $\gamma = 0$.
Upper surface temperatures 4°C or 6°C as indicated.
(After George et al. Copyright 1989 by Gordon and Breach.
Reprinted with permission.)

In the context of patterned ground formation the two important dimensionless numbers are the critical Rayleigh number, $Ra = \xi^3 R^2$, and the critical wave number k. The critical Rayleigh number determines the conditions at the onset of convective motion of water in the soil. The parameters $g, \alpha, \rho_0, \kappa, \mu$ in the Rayleigh number vary little and may be treated as con-

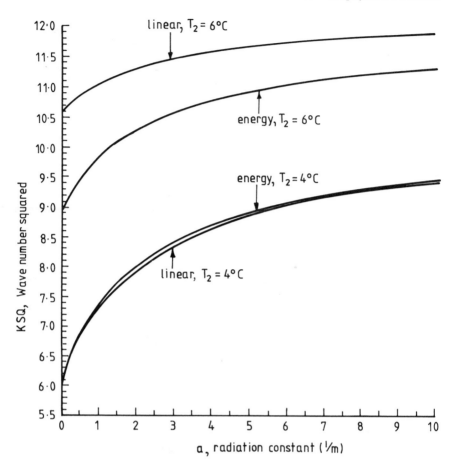

Figure 7.3. The square of the wave number against a, $\gamma = 0$.
Upper surface temperatures 4°C or 6°C as indicated.
(After George et al. Copyright 1989 by Gordon and Breach.
Reprinted with permission.)

stant. The permeability is a separate issue since this changes greatly from
one geographical location to another. Thus, in a sense, the permeability
is the most important parameter in the Rayleigh number affecting stability. Attempts to obtain accurate measurements of soil permeability have
proved extremely difficult. Because the soil in a stone polygon undergoes
substantial lofting due to freeze-thaw cycles, it has proved impossible to
measure accurate permeability values by transporting the soil back to the

laboratory; the soil collapses and the effect of lofting is lost. Moreover, it is not known whether the values of permeability taken *on site* represent the state of the art when the convection process commenced. In this regard field work has not been too reliable and a comparison of Rayleigh numbers to predict the onset of instability conditions is not terribly useful, although many theoretical results are available in George et al. (1989).

The comparison of nonlinear against linear theory is very good. The data presented in George et al. (1989), see also Figures 7.1 to 7.3, demonstrates a maximum difference between the two approaches of less than 20% for the Rayleigh number, and less than 15% for the wave number. They demonstrate that upper surface temperatures greater than 4°C have a stabilizing influence, and it is probably realistic to consider only those upper temperatures in the range 4°C or less.

Wave number predictions have proved to be extremely useful. We take the wave number to measure the width-to-depth of a stone polygon via the equation

$$k = 7.664 \, \frac{d}{L} \, ,$$

where L is the diameter of a single circular convection cell. It is found that the wave number increases with increasing temperatures. Therefore the width-to-depth ratio of stone polygons decreases for higher upper boundary temperatures. The width-to-depth ratios from measurements in the field were found to be higher than the calculated values. This suggests that convective motion in the patterned ground cell begins in the spring as soon as the surface temperature approaches 4°C.

A vertically varying permeability was found theoretically to have almost no effect on the critical wave number. This implies that there should be little variation in the width-to-depth ratio for stone polygons at different sites or when there is a strong vertical variation in permeability. This observation has been borne out by field studies. On the other hand, the theoretical values of the Rayleigh number decrease with a linear increase in permeability, which means the convection threshold is reduced, and this is in accordance with Step 1.

The type of boundary condition that should be used at the upper surface is of importance. Figure 7.3 shows that the width-to-depth values vary by about 30% between the constant temperature and heating solely by radiation cases. The constant temperature condition is consistent with the ground surface being warmed by conduction and convection from the air, and this scenario is likely to be achieved during cloudy or foggy weather or when there are high winds. The constant temperature gradient condition is appropriate when the ground surface is heated directly and evenly by solar radiation. In practice neither condition is likely to be perfect, although it is

believed that radiation heating will be predominant, see e.g., Andersland &
Anderson (1978). The theoretical work shows the constant flux condition is
consistent with a greater width-to-depth ratio and a lower Rayleigh number.
Comparison of the energy and linear results with field data support the use
of the constant flux condition. Indeed, the most realistic case analysed
in George et al. (1989) corresponds to heating by radiation only. The
energy method yields a critical wave number of 2.456. The average width-
to-depth ratio for hexagons is 3.36. George et al. (1989) compare this
theoretical value with a value of 3.61 based on data of Ray et al. (1983) for
a series of sites in the Rocky Mountains and by Gleason et al. (1986) for
additional sites in Iceland and in the Richardson Mountains of the Northern
Yukon, see Tables 7.1, 7.2. The theoretical width-to-depth ratio and the
least squares fit of field data differ by only 7%, which is very good for a
geological phenomenon.

Certainly, the theoretical models of Ray et al. (1983), Gleason et al.
(1986) and George et al. (1989) predict outcomes that are in good agreement with available field studies, and this is very encouraging.

7.4 Convection in Thawing Subsea Permafrost

During the history of the Earth the level of the sea has been dependent on
the size of the glaciers and the quantity of water so held as ice. Around
18000 years ago, see Müller-Beck (1966), p. 1193, the level was some 100–
110m lower than it presently is with the ambient air temperatures being
much colder too, reaching a minimum at roughly the same time, and this led
to substantial permafrost forming around some of the Earth's shores. When
the sea level rose the permafrost responded to the relatively warm and salty
sea, and this has created a thawing front and a layer of salty sediments
beneath the sea bed. Extensive studies of this phenomenon have been
made off the coast of Alaska, see Harrison (1982), Harrison & Osterkamp
(1982), Osterkamp & Harrison (1982), Swift & Harrison (1984), and Swift
et al. (1982). According to Harrison (1982) a very interesting type of
convection is taking place in the layer between the sea bed and permafrost,
the buoyancy mechanism being one of the salty layer melting the relatively
fresh ice, which being less dense then rises through the porous thawing
layer, thereby creating a convective motion.

The mean sea bed temperatures measured by Osterkamp & Harrison
(1982) are around $-1°C$ and a thermal gradient exists with colder temperatures at the downward moving permafrost interface. Harrison (1982)
and Swift & Harrison (1984) employed these facts and by also assuming
the convection is caused by salt effects rather than by temperature were
able to produce a model to describe the subsea convection. The Swift

& Harrison (1984) model is mathematically attractive in that it uses the salt-dominated effect together with the slow, climatic (monotonic) interface advance (2–5cm a^{-1}) to replace a moving boundary problem by one on a fixed region. Hutter & Straughan (1984) show this is correct to leading order on the timescale of convection. It is further interesting to observe that McFadden et al. (1984) employ a similar approximation in crystal growth studies where they argue that, *the onset of instability is approximately the same whether or not the planar crystal melt interface is allowed to deform.* The numerical analysis of Swift & Harrison (1984) treats two-dimensional convection for salt Rayleigh numbers of 1750 and 17500; this is well into the convection regime, with stationary and oscillatory convection values of 40 and 400, see e.g., Caltagirone (1975).

The paper by Galdi et al. (1987) determines the critical Rayleigh number for nonlinear convection by using energy stability theory: the linear instability boundary is obtained as a by-product of the nonlinear analysis. Because the parabolic salt diffusion equation is subject to a nonlinear destabilizing boundary condition at the permafrost interface, that work is unable to preclude the possibility of a large amplitude sub-critical instability. If the Rayleigh number is smaller than a critical one, R_E^α, for convection in a porous layer with a linear flux lower boundary condition then unconditional nonlinear stability holds. In the situation where the nonlinear stability is conditional upon the existence of a finite amplitude threshold, such a threshold value was calculated.

A very interesting mathematical feature of the conditional analysis of Galdi et al. (1987) is that the two- and three-dimensional problems need different analyses. In two dimensions an L^2 energy method employing a Sobolev inequality works well: this method fails for the three-dimensional case, which evidently requires a more subtle approach in which the natural energy has to be generalized to include a controlling term that allows the nonlinearities to be dominated.

Since there has been much recent attention to convection-solidification problems, see e.g., Chorin (1984), Davis et al. (1984), Hurle et al. (1982), McFadden et al. (1984), especially in connection with the Czochralski technique of crystal growing, reviewed at length by Langlois (1985), we believe the nonlinear stability ideas of Galdi et al. (1987), outlined here, may find wider application.

7.5 The Model of Harrison and Swift

The mathematical interpretation of the geological situation is depicted in Figure 7.4. We shall assume, for the purpose of calculating the critical Rayleigh number for the onset of instability, that the permafrost boundary at $z = D(t)$ remains planar. This assumption is reasonable because appreciable boundary movement is on a timescale of years whereas the salt convective motion is on a timescale of days.

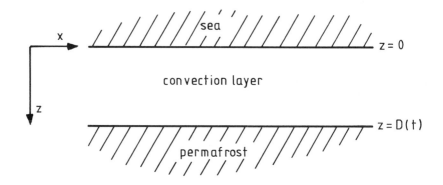

Figure 7.4. Configuration of the thawed layer of salty sediments. (After Galdi et al. (1987).)

The porous layer is modelled by Darcy's law with the body force term linear in the salt field. Employing a Boussinesq approximation the equations of motion are

$$v_i = -\mu p_{,i} + k_i g \mu \rho_0 (1 + S), \tag{7.23}$$

$$v_{i,i} = 0, \tag{7.24}$$

$$\frac{\partial S}{\partial t} + v_i S_{,i} = \kappa \Delta S, \tag{7.25}$$

where (7.25) is the equation for the diffusion of salt in the layer $z \in (0, D)$, $\mathbf{v}, S, \mu, p, \rho_0, \kappa, g$ are velocity, salt concentration, permeability divided by dynamic viscosity, pressure, (constant) density, salt diffusivity, and gravity, respectively, $\mathbf{k} = (0, 0, 1)$, and μ is here taken constant.

It is assumed there is no flow into the sea or permafrost, and so the velocity boundary conditions are

$$v_i n_i = 0, \qquad z = 0, D, \tag{7.26}$$

here n_i denotes the unit outward normal. (Hutter & Straughan (1984) commence with a complete system of equations for velocity, pressure, salt concentration, *and temperature* and exploit the disparity between the timescales of boundary movement and convection. They demonstrate (7.26) is correct at the lowest order in a salt dominated convection régime.) Measurements indicate the salt concentration and temperature at the sea bed are almost constant, and so we choose

$$S = S_0 \text{ (constant)}, \quad z = 0, \tag{7.27}$$

$$T = T_0 \text{ (constant)}, \quad z = 0. \tag{7.28}$$

The lower boundary is moving due to a phase change occurring there; thus, appropriate forms of a Stefan condition must hold there. These are taken to be

$$L\dot{D} = -K_1 \left.\frac{\partial T}{\partial z}\right|_{D-}, \quad z = D, \tag{7.29}$$

$$S(D)\dot{D} = -\kappa \left.\frac{\partial S}{\partial z}\right|_{D-}, \quad z = D, \tag{7.30}$$

where L is the coefficient of latent heat per unit volume of the salty layer (in units of joules m^{-3}), the subscript D^- indicates the derivative approaching D from the thawed layer, K_1 is the thermal conductivity (joules year^{-1} m^{-1} deg^{-1}), with κ the salt diffusivity coefficient (m^2 year^{-1}). The gradients of salt and temperature in the permafrost layer are neglected in (7.29), (7.30).

Swift & Harrison (1984) simplify the above model by arguing that the driving mechanism for convection is the buoyant, relatively pure water melting at $z = D$, and the temperature input may be neglected in the sense that the temperature gradient remains constant throughout; this is in agreement with the measurements of Osterkamp & Harrison (1982). Thus they assume that in the porous layer

$$\frac{\partial T}{\partial z} = \frac{\left[T(D) - T_0\right]}{D}. \tag{7.31}$$

They complete the formulation of the model with a phase equilibrium condition at $z = D$, which with T measured in degrees centigrade is

$$S(D) \propto -T(D). \tag{7.32}$$

From (7.32) we deduce

$$\frac{S(D)}{S_r} = \frac{T(D)}{T_0}, \tag{7.33}$$

where S_r is the salinity of sea water that would begin to freeze at the sea bed temperature T_0. Since $T(D) < T_0 < 0$ it follows that

$$S(D) > S_r.$$

The temperature field is removed from the problem by eliminating $T(D)$ and $\dot{D}$ between (7.29)–(7.31) and (7.33) to yield the nonlinear boundary condition on the salt field,

$$\frac{\partial S}{\partial z} = \frac{K_1 T_0}{L\kappa D} S\left(\frac{S}{S_r} - 1\right), \quad z = D. \tag{7.34}$$

Thus, the governing equations (7.23)–(7.25) are to be solved in conjunction with the boundary conditions (7.26), (7.27), and (7.34), the approximation (7.31) having the effect of fixing the moving boundary at D.

The stationary solution studied in Galdi et al. (1987) is

$$\mathbf{v} \equiv \mathbf{0}, \quad \bar{S} = S_0 - \beta_s z, \quad \bar{p} = p_0 + g\rho_0(1 + S_0)z - \frac{1}{2}g\rho_0\beta_s z^2. \quad (7.35)$$

To investigate the stability of solution (7.35) let $\mathbf{u}$, s, and p be perturbations to the stationary values. Non-dimensionalizing the layer to $\{z \in (0, 1)\}$ and introducing a salt Rayleigh number, Ra, by

$$Ra = R^2 = \frac{\beta_s g\mu\rho_0 D^2}{\kappa},$$

Galdi et al. (1987) derive the following non-dimensional perturbation equations governing the stability of the steady solution (7.35):

$$\begin{aligned} u_i &= -p_{,i} + Rsk_i, \\ u_{i,i} &= 0, \\ s_{,t} + u_i s_{,i} &= Rw + \Delta s, \end{aligned} \quad (7.36)$$

with boundary conditions

$$u_i n_i = 0, \quad z = 0, 1; \quad (7.37)$$

$$s = 0, \quad z = 0; \quad \frac{\partial s}{\partial z} = -as - bs^2 \quad \text{on} \quad z = 1. \quad (7.38)$$

In $(7.36)_3$, $w = u_3$. Moreover, we assume u_i, s, p are periodic functions of x and y with periodicities m and n, respectively. The constants a and b are given by

$$a = \left(2\frac{\bar{S}}{S_r} - 1\right)\frac{K_1|T_0|}{L\kappa}, \quad (7.39)$$

$$b = \frac{K_1|T_0|}{LS_r}\sqrt{\frac{\beta_s}{g\mu\rho_0\kappa}}, \quad (7.40)$$

and we note that estimates of the components in Swift & Harrison (1984) suggest that a will always be positive.

7.6 The Energy Stability Maximum Problem

The stability analysis commences in the usual way. Multiply $(7.36)_1$ by u_i, $(7.36)_3$ by s, add and integrate over a periodicity cell V. Using the boundary conditions and integrating by parts we find

$$\frac{1}{2}\frac{d}{dt}\|s\|^2 = 2R < sw > -\mathcal{D} - b\int_\Gamma s^3 \, dA, \quad (7.41)$$

where we have defined

$$D = \|\mathbf{u}\|^2 + D(s) + a \int_\Gamma s^2 \, dA, \tag{7.42}$$

where $< \cdot >$ and $\| \cdot \|$ denote integration and the norm on $L^2(V)$, respectively, $D(\cdot)$ denotes the Dirichlet integral, and Γ is that part of the boundary of V that lies in the plane $z = 1$.

Define further

$$I = 2 < sw >, \tag{7.43}$$

to derive from (7.41)

$$\frac{1}{2}\frac{d}{dt}\|s\|^2 \leq -\left(\frac{R_E - R}{R_E}\right)D - b\int_\Gamma s^3 \, dA, \tag{7.44}$$

where R_E is defined by

$$\frac{1}{R_E} = \max_{\mathcal{H}} \frac{I}{D}, \tag{7.45}$$

and where $\mathcal{H}$ is the space of admissible functions, which we here choose as

$$\mathcal{H} = \{\mathbf{u}, s | \mathbf{u} \in (L^2(V))^3, \; u_{i,i} = 0, \; u_i n_i = 0 \text{ on } z = 0, 1;$$
$$s \in W^{1,2}(V), \; s = 0 \text{ at } z = 0; \; u_i, s \text{ periodic in } x \text{ and } y \} \tag{7.46}$$

The first result of Galdi et al. (1987) is an *unconditional* one and shows that if

$$R < R_E^\alpha \equiv \max_{\mathcal{H}} \frac{I}{D_\alpha}$$

with D_α defined by

$$D_\alpha \equiv \|\mathbf{u}\|^2 + D(s) + \alpha \int_\Gamma s^2 \, dA,$$

then $\|s(t)\|^2$ decays at least exponentially for increasing t, for all $s(\mathbf{x}, 0) \in L^2(V)$. The constant α is defined by

$$\alpha = \frac{K_1 |T_0|}{L\kappa} \left(\frac{\bar{S}(D)}{S_r} - 1\right).$$

This result hinges on the facts that $\bar{S}(D) > S_r$ and $s + \bar{S}(D) > 0$, i.e., the salinity is always positive. Using these facts they deduce from (7.44)

$$\frac{1}{2}\frac{d}{dt}\|s\|^2 \leq -D_\alpha\left(1 - \frac{R}{R_E^\alpha}\right),$$

and this together with use of Poincaré's inequality yields rapid decay of $\|s(t)\|$ provided $R < R_E^\alpha$.

Decay for the case $R_E^\alpha < R < R_E$ is trickier and requires use of embedding inequalities. Interestingly, the two- and three-dimensional problems evidently need separate treatments. This is described in §7.7.

An existence result for the maximizing solution to (7.45) (or the equivalent problem involving $\mathcal{D}_\alpha$) is proven in Galdi et al. (1987). To achieve this they observe that $\mathcal{H}$ is a Hilbert space when equipped with the norm generated by $\mathcal{D}$ (where the boundary conditions are to be interpreted in the trace sense). To see this, observe that $\mathcal{H}$ is the topological product of two complete spaces and is hence itself complete with respect to the norm endowed by $\mathcal{D}$. The completeness of the space appropriate to **u** follows from, for example, Temam (1978), whereas the space appropriate to s is a subspace of $W^{1,2}(V)$ endowed with the standard norm; this follows because

$$\|s\|^2 + D(s) \leq (1 + 4\pi^{-2})D(s)$$

$$\leq (1 + 4\pi^{-2})\left(D(s) + a\int_\Gamma s^2\, dA\right), \qquad (7.47)$$

$$\leq (1 + 4\pi^{-2})\left[D(s) + 2a\|s\|D^{1/2}(s)\right]$$

$$\leq (1 + 4\pi^{-2})(1 + a)\left[D(s) + \|s\|^2\right], \qquad (7.48)$$

where the inequality involving the boundary integral may be found in Appendix 1.

Thanks to Poincaré's inequality $I/\mathcal{D}$ is bounded above by a constant, γ, say. Hence, there exists a maximizing sequence $\{\mathbf{u}_n, s_n\}$ such that

$$\lim_{n\to\infty} \frac{I(w_n, s_n)}{\mathcal{D}(\mathbf{u}_n, s_n)} = \gamma.$$

This sequence is chosen such that $\mathcal{D}(\mathbf{u}_n, s_n) = 1$ and from this we may deduce the existence of a subsequence, again denoted by $\{\mathbf{u}_n, s_n\}$, such that

$$u_{i_n} \to u_{i_0} \quad \text{weakly in} \quad L^2(V), \qquad (7.49)$$

$$s_n \to s_0 \quad \text{weakly in} \quad W^{1,2}(V), \quad \text{strongly in} \quad L^2(V), \qquad (7.50)$$

for some $\{\mathbf{u}_0, s_0\} \in \mathcal{H}$. By using standard inequalities,

$$|I(w_n, s_n) - I(w_0, s_0)| \leq 2|< s_0(w_n - w_0) >| + 2|< w_n(s_n - s_0) >|$$

$$\leq 2|< s_0(w_n - w_0) >| + 2\|w_n\|\,\|s_n - s_0\|. (7.51)$$

The first term on the right of (7.51) converges to zero by virtue of (7.49), while the second tends to zero thanks to (7.50). Therefore, $I(w_0, s_0) = \gamma$ and also one may show $\mathcal{D}(\mathbf{u}_0, s_0) = 1$. The existence of a maximizing solution to (7.45) is therefore established.

To actually solve (7.45) we find the Euler-Lagrange equations to be

$$R_E s k_i + p_{,i} = u_i, \qquad (7.52)$$

$$R_E w + \Delta s = 0, \qquad (7.53)$$

together with the boundary conditions

$$u_i n_i = 0, \quad z = 0, 1; \tag{7.54}$$

$$s = 0, \quad z = 0; \qquad \frac{\partial s}{\partial z} + as = 0 \quad \text{on} \quad z = 1. \tag{7.55}$$

It is very important to observe that (7.52)–(7.55) are *exactly the same as the linearized version of the full equations* (7.36)–(7.38). It transpires that this means that the critical Rayleigh number is the *same for both linear and nonlinear (conditional) stability.* Any finite amplitude instability is caused entirely by the bs^2 term in (7.38).

The critical value of R_E is found from (7.52)–(7.55) by a numerical technique, c.f. Appendix 2. Figure 7.5 presents graphical output of the Rayleigh number against a for $a \leq 100$; in the limit $a \to \infty$, Ra is asymptotic to $4\pi^2$.

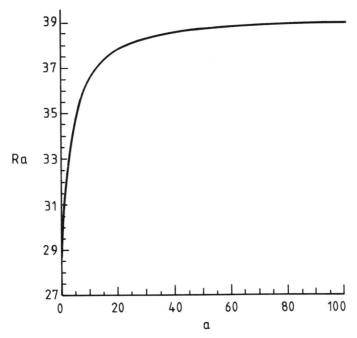

Figure 7.5. Critical values of Ra against a.
(After Galdi et al. (1987).)

7.7 Decay of the Energy

A nonlinear stability result follows directly from (7.44) if we consider only *two-dimensional perturbations* and employ inequality (7.56), namely,

$$\int_\Gamma s^3 \, dA \leq c_1 \|s\| D(s), \tag{7.56}$$

see Appendix 1. Suppose now,

$$R < R_E, \tag{7.57}$$

then with $h = R^{-1} - R_E^{-1}$ (> 0), (7.44) yields

$$\frac{1}{2}\frac{d}{dt}\|s\|^2 \leq -D(s)(hR - bc_1\|s\|) - hR\left(\|\mathbf{u}\|^2 + a\int_\Gamma s^2 \, dA\right). \tag{7.58}$$

Suppose further that the initial data satisfy

$$\|s(0)\| < \frac{R_E - R}{R_E c_1 b}. \tag{7.59}$$

Then, with

$$\omega = hR - bc_1\|s(0)\|,$$

we may deduce from (7.58) that

$$\frac{1}{2}\frac{d}{dt}\|s\|^2 + \frac{1}{4}\pi^2\omega\|s\|^2 + hR\left(\|\mathbf{u}\|^2 + a\int_\Gamma s^2 \, dA\right) \leq 0. \tag{7.60}$$

Dropping the last term and integrating we arrive at

$$\|s(t)\|^2 \leq \|s(0)\|^2 e^{-\frac{1}{2}\pi^2\omega t}. \tag{7.61}$$

Since from (7.36)$_1$,

$$\|\mathbf{u}\|^2 = R < sw >,$$

it is easy to see that

$$\|\mathbf{u}\|^2 \leq \frac{4}{3}R^2\|s\|^2,$$

and this in (7.61) implies also decay of $\|\mathbf{u}\|^2$.

Nonlinear stability of the solution (7.35) to two-dimensional perturbations is assured if

$$\text{(i)} \quad R < R_E; \quad \text{(ii)} \quad \|s(0)\| < \frac{hR}{3b(2 + 4/\pi m)^{1/2}}, \tag{7.62}$$

where m is the x-periodicity of u_i, s. Condition (i) determines the nonlinear critical Rayleigh number whereas (ii) is a limitation on the size of the initial amplitude.

Galdi et al. (1987) show that inequality (7.56) is false in three dimensions (see Appendix 1). Hence the above procedure to establish conditional decay fails when three-dimensional perturbations are taken into consideration. To retain the feature that the linear and nonlinear Rayleigh number boundaries are the same, the outlet adopted by Galdi et al. (1987) is to use the generalized energy

$$E(t) = \frac{1}{2}\|s\|^2 + \frac{1}{4}\mu\|s^2\|^2, \tag{7.63}$$

for $\mu\,(>0)$ a coupling parameter. In (7.63) the L^4 piece is added for the same motive as similar additions in chapter 6, that is wholly to control the cubic nonlinear terms, and it plays no role in determining the stability boundary.

With $m = R/R_E$ and $R < R_E$, the energy inequality for E defined by (7.63) is shown by Galdi et al. (1987) to satisfy

$$\frac{dE}{dt} \leq -\mathcal{D}(1-m) - b\int_\Gamma s^3 dA + \mu R < s^3 w >$$
$$- \mu a \int_\Gamma s^4 dA - \mu b \int_\Gamma s^5 dA - \frac{3}{4}\mu D(s^2)$$
$$- \lambda D(w) + \lambda R < s_{,\alpha} w_{,\alpha} >, \tag{7.64}$$

where $\lambda\,(>0)$ is another coupling parameter and the subscript α sums over 1 and 2. The coupling parameter λ, like μ, is employed only to control and dominate the nonlinear terms; the energy stability boundary arises through the L^2 energy.

The addition of the L^4 integral of s to the energy E in (7.63) adds higher order dissipative terms to (7.64), which now allows the nonlinear boundary terms to be controlled and conditional energy decay is again achieved. The nonlinear stability criterion is still $R < R_E$, with $E(0)$ suitably small. It is, however, interesting to observe that the generalized energy (7.63) *must* be employed (as opposed to simply the L^2 energy) to achieve energy decay in three dimensions.

The problem of convection in thawing subsea permafrost is further investigated using energy theory by Payne et al. (1988). These writers take into account temperature variations and their analysis employs a weighted energy. Hutter & Straughan (1984) employ weakly nonlinear and energy techniques and they find the permafrost boundary has a parabolic shape approaching the shoreline.

8

Surface Tension Driven Convection

8.1 Energy Stability and Surface Tension Driven Convection

The topic of this section is historically important since it is now believed that in Bénard's (1900) original experiments on the convection problem, which now bears his name, the driving mechanism was one of surface tension variation due to temperature. We point out that convective instability seems to have been first described by James Thomson (1882), the elder brother of Lord Kelvin: a very clear account of this is given in Drazin & Reid (1981), p. 32.

The first analysis of surface tension driven convection by means of the energy method is due to Davis (1969b). He assumes a flat surface, a plane layer, and a linear surface tension temperature relation of the form

$$\sigma = \sigma_0 - \gamma(T - T_0), \tag{8.1}$$

σ, T being surface tension and temperature and σ_0, γ positive constants. His non-dimensional perturbation equations are

$$Pr^{-1}\left(\frac{\partial u_i}{\partial t} + u_j u_{i,j}\right) = -p_{,i} + \Delta u_i + R\theta k_i,$$

$$\frac{\partial \theta}{\partial t} + u_i \theta_{,i} = \Delta\theta + w, \qquad u_{i,i} = 0,$$

where Pr is the Prandtl number, other variables being as before. The boundary conditions examined by Davis (1969b) are

$$u_i = \theta = 0, \qquad z = 0,$$
$$\theta_z + L\theta = w = u_z + B\theta_x = v_z + B\theta_y = 0, \qquad z = 1. \tag{8.2}$$

In (8.2), $(u_1, u_2, u_3) = (u, v, w)$, a subscript $x, y,$ or z denotes differentiation with respect to that variable, L is a positive constant, B is the Marangoni number, and the last two conditions represent the fact that the shear stress in the surface is given by a surface tension gradient.

Let V be a period cell for the motion. Davis (1969b) *claims* that the B terms in (8.2) act on the system like a bounded operator. He shows that for a periodic disturbance

$$\int_\Gamma (\phi_x^2 + \phi_y^2)\, dA = a^2 \int_\Gamma \phi^2\, dA,$$

where $\phi = R^{1/2}\theta$, Γ is that part of $z = 1$ intersecting ∂V, and a is the wave number, and then concludes that the B terms act like a bounded operator. His argument is based on the fact that the numerical results indicate a^2 remains bounded so he deduces (p. 351): *By restricting ourselves to a finite interval of wave-numbers, due to a physical preference, we see that...behaves as would a bounded perturbation operator.* His dissipation term contains the square of the gradients of ϕ integrated over V and so it is possible to dominate the surface terms, but only by the Laplacian, i.e., the unbounded operator. Therefore, from a mathematical point of view the claim that the B terms act like a bounded operator is very strong. His energy theory chooses

$$E = \frac{1}{2}(Pr^{-1}\|\mathbf{u}\|^2 + \lambda\|\theta\|^2),$$

for $\lambda\,(> 0)$ a coupling parameter. Then he derives

$$\frac{dE}{dt} = (R + \lambda) < w\theta > -\,[D(\mathbf{u}) + \lambda D(\theta)] - B\int_\Gamma \theta w_z\, dA - L\int_\Gamma \theta^2\, dA.$$

A symmetrization is introduced by setting $\phi = \lambda^{1/2}\theta$, $\lambda = B\mu$, $R = BN_r$, and from this he uses the variational theory of energy stability to deduce that $E \to 0$, $t \to \infty$, provided

$$B < \rho^2,$$

where ρ is defined by

$$\frac{1}{\rho} = \max_{\mathcal{H}}\left\{\frac{\mu + N_r}{\sqrt{\mu}} < w\phi > +\frac{1}{\sqrt{\mu}}\int_\Gamma (-\phi w_z)\, dA\right\}, \qquad (8.3)$$

with u_i, ϕ constrained by the relation

$$D(\mathbf{u}) + D(\phi) + L\int_\Gamma \phi^2\, dA = 1.$$

The solution to (8.3) is completed by deriving the Euler-Lagrange equations in the form

$$\Delta\phi + \frac{1}{2}B_\mu\,\frac{\mu + N_r}{\sqrt{\mu}}\,w = 0, \qquad (8.4)$$

$$\Delta u_i + \frac{1}{2}B_\mu\,\frac{\mu + N_r}{\sqrt{\mu}}\,\phi k_i = \pi_{,i}, \qquad (8.5)$$

$$u_{i,i} = 0, \qquad (8.6)$$

in $\{z \in (0,1)\} \times \mathbf{R}^2$, together with the boundary conditions

$$u_i = \phi = 0, \qquad z = 0, \tag{8.7}$$

$$\phi_z + L\phi + \frac{1}{2}\frac{B_\mu}{\sqrt{\mu}}\, w_z = u_z + \frac{1}{2}\frac{B_\mu}{\sqrt{\mu}}\, \phi_x$$

$$= v_z + \frac{1}{2}\frac{B_\mu}{\sqrt{\mu}}\, \phi_y = w = 0, \qquad z = 1. \tag{8.8}$$

Davis (1969b) also derives several useful relations by parametric differentiation. He solves (8.4)–(8.8) by introducing a wave number in the x, y directions and integrating the resulting one-dimensional eigenvalue problem by using a shooting method. He compares his results against those of linear theory and shows they are very sharp, e.g., when $L = B = 0$, $R_L = R_E = 669.0$, and even when $L = 0$, $R_E, R_L \to 0$, the respective B values are still relatively close, being $B_E = 56.77$, $B_L = 79.61$. Thus, even for surface tension driven convection the energy method proves very useful.

   There are many generalizations and extensions of Davis' (1969b) work. For example, Davis & Homsy (1980) extend the analysis to the situation where the surface is allowed to deform; this is a very technical piece of work and they find power series solutions to the Euler-Lagrange equations in the Crispation number (which measures surface deflection). Castillo & Velarde (1983) develop an equivalent analysis when a concentration field, C, is also present and when $\sigma = \sigma(T, C)$. McTaggart (1983a,1983b) also considers the linear problem when $\sigma = \sigma(T, C)$ for an infinite plane layer, and for a bounded geometry. She shows that in the limit of zero buoyancy convection when both the Marangoni and solutal Marangoni numbers are positive, oscillatory instability does not occur; when, however, the Marangoni numbers have opposite signs oscillatory instability occurs. Computations are performed for an aqueous solution of magnesium sulphate. Other interesting results on surface tension convection in bounded geometries are contained in Rosenblat, Davis & Homsy (1982) and Rosenblat, Homsy & Davis (1982); these and other aspects are reviewed by Davis (1987).

8.2 Surface Film Driven Convection

Davis & Homsy (1980) also briefly consider an ad hoc theory where the surface is itself a two-dimensional continuum. A more complete approach to this problem is by McTaggart (1983a,1984) who employs the (non ad hoc) surface theory, based on a rigorous development from continuum thermodynamics, of Lindsay & Straughan (1979). Since the study of McTaggart is relevant to the technologically important problem of convection where a thin film of possibly different fluid overlies the convecting layer we include a description of her work.

It is known from experiments, see e.g., Berg & Acrivos (1965), that the presence of a surface film, overlying a layer of fluid heated from below, has a pronounced effect on the onset of convection. Results pertaining to the onset of such convection are important, for example, in the field of crystal growth where convective motion within the liquid during solidification can result in crystal defects, see e.g., Antar et al. (1980).

The work of McTaggart (1983a,1984) extends that of Berg & Acrivos (1965) who examined the effect of surface active agents on convection cells induced by surface tension in shallow layers of fluid. McTaggart examines the more realistic situation where the bulk fluid is not necessarily shallow, by adopting an alternative approach to the introduction of surface tension. The film is regarded as a two-dimensional continuum and surface tension is then introduced naturally as a combination of a surface density and the derivative of a surface free energy.

The model adopted originated with work of Landau & Lifschitz (1959), p. 241, on the effect of adsorbed films on the motion of a liquid. Scriven (1960) extended these ideas and developed the momentum equations for a two-dimensional continuum and this approach was fully developed according to the methods of modern continuum thermodynamics by Lindsay & Straughan (1979).

An advantage of the model of Lindsay & Straughan (1979) is that it enables one not only to examine the role of interfacial tension, but also the part played by other interfacial properties that pertain to the resistance of an interface to deformation. In this way McTaggart (1983a,1984) is able to measure the stabilizing effect of the viscosity and the thermal conductivity of the film on the onset of convection cells induced by buoyancy forces.

The mathematical picture is now described. A thin film of fluid overlies a layer of different (immiscible) fluid, which is referred to as the bulk fluid. The bulk fluid is contained between the planes $z = 0$ and $z = d \ (> 0)$ and is subjected to heating from below. The equations for the bulk fluid are

$$
\begin{aligned}
v_{i,i} &= 0, \\
\rho \dot{v}_i &= -p_{,i} + \mu \Delta v_i + \rho g_i \big[1 - \alpha(T - T_R)\big], \\
\dot{T} &= \kappa \Delta T,
\end{aligned}
\qquad (8.9)
$$

where v_i is the velocity in the bulk fluid, T is temperature, ρ density, p pressure, μ is the viscosity, $\mathbf{g} = (0, 0, -g)$, g being the magnitude of the acceleration due to gravity, α is the coefficient of thermal expansion, T_R is a reference temperature, and κ is the thermal diffusivity.

The film is regarded as an interface between the air (region V^+) and the bulk fluid (region V^-). Assuming the interface to be flat, equations (2.9) of Lindsay & Straughan (1979), being the equations of mass, momentum, and energy in the surface, reduce to

$$\dot{\gamma} + \gamma V^{\alpha}{}_{;\alpha} = 0,$$
$$\gamma \dot{V}^{k} - S^{k\alpha}{}_{;\alpha} = [t^{ki} n^{i}] + \gamma f^{k}, \tag{8.10}$$
$$\gamma \dot{\epsilon} + q^{\alpha}{}_{;\alpha} - S^{k\alpha} V^{k}{}_{;\alpha} - \gamma r = -[Q^{i} n^{i} + t^{ki} n^{k} (\dot{x}^{i} - \dot{\xi}^{i})],$$

where γ is the surface density, $\dot{x}^{i} = V^{i}$ is the velocity of a material particle in the surface and V^{α} are the tangential components of V^{i}, $S^{k\alpha}$ is the surface stress, t^{ik} is the Cauchy stress, n_{i} denotes the unit outward normal, f^{i} is the surface body force, ϵ is the specific internal surface energy, q^{α} and Q^{i} are the surface and bulk heat flux vectors, respectively, r is the surface heat supply, and $\dot{\xi}^{i} = v^{i}$. A superposed dot denotes material differentiation, a subscript semicolon denotes differentiation with respect to the surface variables, $x, y,$ and $[\phi] = \phi^{+} - \phi^{-}$ denotes the jump in a quantity across the interface. The summation convention is employed with repeated Greek letters indicating summation from 1 to 2 while repeated Latin letters take values from 1 to 3. Since the interface is bounded by an inviscid fluid above and a viscous fluid below, it follows from the Clausius-Duhem inequality that

$$[t^{ki} n^{k} (\dot{x}^{i} - \dot{\xi}^{i})] = 0.$$

It is additionally assumed that $f^{k} \equiv 0$.

By analogy with three-dimensional theories, a surface free energy ψ is introduced by

$$\psi = \epsilon - \eta \bar{\theta},$$

where $\bar{\theta}$ and η are the temperature and specific entropy in the film and ψ, η satisfy (Lindsay & Straughan (1979))

$$\psi = \psi(\bar{\theta}, \gamma), \qquad \eta = -\frac{\partial \psi}{\partial \bar{\theta}}.$$

A constitutive theory is then proposed for which ψ has form

$$\psi = C_{A} \bar{\theta} \left[1 - \log \left(\frac{\bar{\theta}}{\bar{\theta}_{R}} \right) \right] - \frac{s}{\gamma} (\bar{\theta} - \bar{\theta}_{R}),$$

where $\bar{\theta}_{R}$ is a reference temperature, C_{A} is the specific heat per unit area, and s is a positive constant. Equation (5.4) of Lindsay & Straughan (1979) then shows the surface tension σ has the form

$$\sigma = -\gamma^{2} \frac{\partial \psi}{\partial \gamma} = -s(\bar{\theta} - \bar{\theta}_{R});$$

thus it satisfies the (often assumed) condition that σ be a linear, decreasing function of $\bar{\theta}$.

The appropriate representations for $S^{\alpha\beta}$ and q^α are

$$S^{\alpha\beta} = \sigma a^{\alpha\beta} + \nu_2 a^{\alpha\beta} d^\mu_\mu + \nu_6 d^{\alpha\beta},$$

$$q^\alpha = q_0 g^\alpha,$$

with $d_{\alpha\beta} = \frac{1}{2}(V_{\alpha;\beta} + V_{\beta;\alpha})$, $g_\alpha = \bar{\theta}_{;\alpha}$, $S^{\alpha\beta}x^k_{;\beta} = S^{k\alpha}$, and $a_{\alpha\beta} = x^i_{;\alpha}x^i_{;\beta}$.
Moreover, it is assumed that ν_2, ν_6, the coefficients of surface dilational and surface shear viscosity, and q_0, the thermal conductivity of the film, are constant. From the entropy inequality, it follows that $q_0 < 0$. McTaggart (1983a,1984) observes that techniques for measuring ν_2 and ν_6 are available in Briley et al. (1976). Since the bulk fluid is viscous, $v^i = V^i$ on the surface, and the temperature is assumed continuous between the bulk fluid and the film so $\bar{\theta} = T$ on $z = d$. For the assumption of a flat interface it may then be deduced that

$$v^i_{;\alpha} = V^i_{;\alpha} \quad \text{and} \quad g_\alpha = \bar{\theta}_{;\alpha} = T_{;\alpha} \quad \text{on} \quad z = d.$$

The film equations (8.10) are now rewritten

$$\gamma\dot{v}^k = -sT_{;\alpha}x^k_{;\alpha} + \nu_2 d^\mu_{\mu;\alpha}x^k_{;\alpha} + \nu_6 d^{\alpha\beta}_{;\alpha}x^k_{;\beta} - 2\mu d^{3k},$$

$$\gamma C_v \dot{T} = -q_0 g^\alpha_{;\alpha} - sTv^\alpha_{;\alpha} - k\frac{\partial T}{\partial z} + \gamma r, \tag{8.11}$$

where $d_{ij} = \frac{1}{2}(v_{i,j} + v_{j,i})$ and k is the thermal conductivity of the bulk fluid. In the derivation of (8.11) terms of second order in the velocity gradients are neglected, a procedure consistent with the derivation of $(8.9)_3$.

The steady state solution (indicated by a superposed hat) to equations (8.9) and (8.11), satisfying the boundary conditions $T(0) = T_0$ and $T(d) = T_d$, is

$$\hat{v}_i = 0, \quad \hat{T} = T_0 - \beta z, \quad \hat{r} = -\frac{k\beta}{\gamma},$$

$$\hat{p} = p_0 - \rho g z + \alpha\rho g\left[(T_0 - T_R)z - \frac{1}{2}\beta z^2\right],$$

where $\beta = (T_0 - T_d)/d$ and p_0 is a constant.

To study the stability of the above steady state solution she considers a perturbation (u_i, θ, π) to $(\hat{v}_i, \hat{T}, \hat{p})$. The resulting non-dimensional governing equations are written as:

Bulk:

$$u_{i,i} = 0,$$

$$\dot{u}_i = -\pi_{,i} + \Delta u_i + R\theta\delta_{i3}, \tag{8.12}$$

$$Pr\dot{\theta} = Rw + \Delta\theta;$$

Film:

$$S_1\dot{u}^k = -\frac{B}{R}\theta_{;\alpha}x^k_{;\alpha} + A_1 d^\mu_{\mu;\alpha}x^k_{;\alpha} + A_2 d^{\alpha\beta}_{;\alpha}x^k_{;\beta} - 2d^{3k},$$

$$S_1 Pr\dot{\theta} = A_3 g^\alpha_{;\alpha} - \frac{B}{R}A_4 u^\alpha_{;\alpha} - A_5\theta u^\alpha_{;\alpha} - \frac{\partial\theta}{\partial z}, \tag{8.13}$$

where $(u_1, u_2, u_3) = (u, v, w)$ and the Rayleigh, Prandtl, and Marangoni numbers are defined by

$$R^2 = \frac{g\alpha\beta d^4}{\kappa\nu}, \qquad Pr = \frac{\nu}{\kappa}, \qquad B = \frac{s\beta d^2}{\rho\kappa\nu}.$$

The non-dimensional numbers $S_1, A_1, \ldots, A_5$ are defined by

$$S_1 = \frac{\gamma}{\rho d}, \qquad A_1 = \frac{\nu_2}{\mu d}, \qquad A_2 = \frac{\nu_6}{\mu d},$$

$$A_3 = -\frac{q_0}{dk}, \qquad A_4 = \frac{T_d g\alpha}{C_v\beta}, \qquad A_5 = \frac{sd}{\rho\kappa\nu}.$$

The perturbations are taken to be periodic in x and y. A typical period cell in the non-dimensional layer is denoted by V, and that part of the surface $z = 1$ that forms part of the boundary of V is denoted by Γ. The lower boundary is rigid and held at constant temperature, so

$$u_i = \theta = 0, \qquad \text{on} \qquad z = 0.$$

To study stability McTaggart (1983a,1984) defines an energy $E(t)$ by

$$E(t) = \frac{1}{2}\int_V (u^i u^i + Pr\theta^2)dx + \frac{1}{2}\int_\Gamma (u^i u^i + Pr\theta^2)da.$$

Then using (8.12), (8.13) it is found that

$$\begin{aligned}
\frac{dE}{dt} = &- D(\mathbf{u}) - D(\theta) + 2R < \theta w > -\frac{B}{R}(1 - A_4)\int_\Gamma \theta_{;\alpha} u^\alpha \, da \\
&- A_3 \int_\Gamma \theta_{;\alpha}\theta_{;\alpha} \, da - A_1 \int_\Gamma (d_\mu^\mu)^2 da \\
&- A_2 \int_\Gamma d^{\alpha\beta}d_{\alpha\beta}da - A_5 \int_\Gamma \theta^2 u^\alpha_{;\alpha}da,
\end{aligned} \tag{8.14}$$

where $< \cdot >$ and $D(\cdot)$ denote integration over V and the Dirichlet integral on V. McTaggart (1983a,1984) introduces the change of variables $B = R^2 N$ and $A_5 = R\bar{T}$, and then shows that

$$\frac{dE}{dt} = -\mathcal{D} + RI \le -\mathcal{D}\left[1 - R \max \frac{I}{\mathcal{D}}\right],$$

where

$$\mathcal{D} = D(\mathbf{u}) + D(\theta) + A_3 \int_\Gamma \theta_{;\alpha}\theta_{;\alpha} \, da + A_1 \int_\Gamma (d_\mu^\mu)^2 da + A_2 \int_\Gamma d^{\alpha\beta}d_{\alpha\beta}da,$$

$$I = 2 < \theta w > -N(1 - A_4)\int_\Gamma \theta_{;\alpha}V^\alpha da - \bar{T}\int_\Gamma \theta^2 V^\alpha_{;\alpha}da, \tag{8.15}$$

and where the maximum is over the space of admissible functions. She writes

$$\frac{1}{\Lambda} = \max \frac{I}{\mathcal{D}}, \tag{8.16}$$

and then argues provided $R < \Lambda$ one may employ Poincaré's inequality to deduce that, for a positive constant k,

$$\frac{dE}{dt} \le -kE\left(1 - \frac{R}{\Lambda}\right);$$

hence global stability follows. The nonlinear stability problem, therefore, reduces to solving the maximum problem (8.16). The Euler-Lagrange equations corresponding to (8.16) are

$$\begin{aligned} R_E w + \Delta\theta &= 0, && x \in V, \\ R_E \theta \delta_{j3} + \Delta u_j - P_{,j} &= 0, && x \in V, \end{aligned} \tag{8.17}$$

where R_E and $P(x, y, z)$ are Lagrange multipliers, with the natural boundary conditions

$$u_i = \theta = 0, \qquad \text{on} \qquad z = 0; \tag{8.18}_1$$

$w = 0,$

$$\frac{\partial\theta}{\partial z} + \bar{T}R_E\theta u^\alpha{}_{;\alpha} - \frac{1}{2}R_E N(1 - A_4)u^\alpha{}_{;\alpha} - A_3\theta_{;\alpha\alpha} = 0,$$

$$\frac{\partial u}{\partial z} + R_E\left(\frac{N}{2} - \frac{A_4}{2} - \bar{T}\theta\right)\frac{\partial\theta}{\partial x} - A_1\frac{\partial}{\partial x}d^\mu_\mu - A_2 d^{1\beta}{}_{;\beta} = 0, \tag{8.18}_{2-5}$$

$$\frac{\partial v}{\partial z} + R_E\left(\frac{N}{2} - \frac{A_4}{2} - \bar{T}\theta\right)\frac{\partial\theta}{\partial y} - A_1\frac{\partial}{\partial y}d^\mu_\mu - A_2 d^{2\beta}{}_{;\beta} = 0,$$

where $(8.18)_{2-5}$ hold on the surface $z = 1$.

   McTaggart (1983a,1984) notes that (8.17), (8.18) form a *nonlinear* system; the bulk equations (8.17) being linear, but the natural boundary conditions nonlinear. She then observes that it may be possible to solve this system using a perturbation expansion technique as in Davis & Homsy (1980), although she only considers the linearized version. In so doing she obtains, to first order (in $\bar{T}$) the energy limit R_E. Such a step, she argues, is a necessary preliminary to the study of the nonlinear system, and so she proceeds by neglecting the nonlinear term $\int_\Gamma \theta^2 V^\alpha{}_{;\alpha} da$. The natural boundary conditions are hence linear and reduce to

$$u_i = \theta = 0, \qquad \text{on} \qquad z = 0,$$

$$w = w_{zz} - (A_1 + A_2)\Delta^* w_z$$

$$\qquad - \frac{1}{2}R_E N(1 - A_4)\Delta^*\theta = 0, \qquad \text{on} \quad z = 1, \tag{8.19}$$

$$\theta_z + \frac{1}{2}R_E N(1 - A_4)w_z - A_3\Delta^*\theta = 0, \qquad \text{on} \qquad z = 1,$$

with $\Delta^* \equiv \partial^2/\partial x^2 + \partial^2/\partial y^2$. It may now be verified that $R_E = \Lambda$ and hence $R < R_E$ guarantees global stability. The work of McTaggart (1983a,1984) then carries out a numerical evaluation of R_E for various choices of parameters.

The maximum problem of (8.16) is not well set since the numerator contains a cubic term on the boundary. Hence McTaggart has essentially to remove this term by choosing $A_5 = 0$. Of course this is a strong requirement; since $A_5 \propto sd$ it means either surface tension is neglected or vanishingly thin layer theory is studied. Therefore, we now indicate another approach that allows non-zero A_5. Instead of defining I as in (8.15) we omit the cubic term and hence define

$$I = 2 < \theta w > -N(1 - A_4) \int_\Gamma \theta_{;\alpha} V^\alpha da.$$

This then yields a well-defined maximum problem that, in fact, reduces to equations (8.17) and (8.19), and hence the results for R_E derived by McTaggart continue to hold. To handle the nonlinearity, however, we observe that by use of the Cauchy-Schwarz inequality

$$\int_\Gamma \theta^2 V^\alpha_{;\alpha} da \le \left(\int_\Gamma \theta^4 da \right)^{1/2} \left(\int_\Gamma (d^\mu_\mu)^2 da \right)^{1/2}. \tag{8.20}$$

The d^μ_μ term is contained in $\mathcal{D}$ and to estimate the θ^4 term we additionally suppose $\int_\Gamma \theta \, da = 0$ and then use the surface Sobolev inequality (Appendix 1),

$$\left(\int_\Gamma \theta^4 da \right)^{1/2} \le c \left(\int_\Gamma \theta_{;\alpha}\theta_{;\alpha} da \right)^{1/2} \left(\int_\Gamma \theta^2 da \right)^{1/2}.$$

Employing this in the energy equation (8.14) we may derive

$$\frac{dE}{dt} \le -\mathcal{D}\left(\frac{R_E - R}{R_E} \right) + c\bar{T}\left(\frac{2}{S_1 Pr A_1 A_3} \right)^{1/2} E\mathcal{D}.$$

Hence, provided

$$R < R_E \quad \text{and} \quad E(0) < \left(\frac{R_E - R}{R_E c\bar{T}} \right) \left(\frac{S_1 Pr A_1 A_3}{2} \right)^{1/2},$$

a standard argument employing Poincaré's and Wirtinger's inequality (cf. chapter 4) shows there is at least exponential decay of $E(t)$ and hence global, conditional stability. Thus, McTaggart's (1983a,1984) results now represent a rigorous and complete nonlinear stability theory.

We point out that in the manipulation of the nonlinearity it is essential that the energy contains surface temperature terms. An attempt to dominate the nonlinearity in (8.20) by bulk terms (as is done in the thawing subsea permafrost problem in chapter 7) evidently fails here. Joseph (1988) also finds difficulty in applying energy theory to convection driven by interfacial tension between two fluid layers. It would be interesting to see if a thin film theory will overcome this difficulty.

8.3 Conclusions from Surface Film Driven Convection

The conclusions McTaggart (1983a,1984) draws from her analysis are worth noting. She examines how the effects of surface viscosity, $V = A_1 + A_2$, surface thermal conductivity, A_3, and surface tension, as measured by $S = N(1 - A_4)$, each affect the onset of convection in the bulk fluid. Figure 8.1 plots the stability curve of energy theory for increasing thermal conductivity A_3, and for zero surface viscosity. In Figure 8.2 the stability curves are shown for several (fixed) values of A_3 with increasing surface viscosity. In both Figures 8.1 and 8.2, $s = 0$.

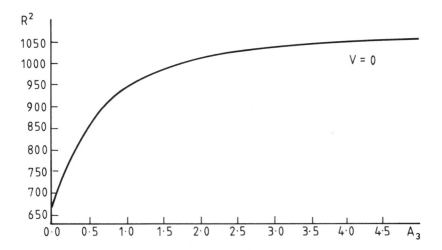

Figure 8.1. Energy stability threshold; Rayleigh number, R^2, against thermal conductivity, A_3.
(After McTaggart. Copyright 1984 by Pergamon Press PLC. Reprinted with permission.)

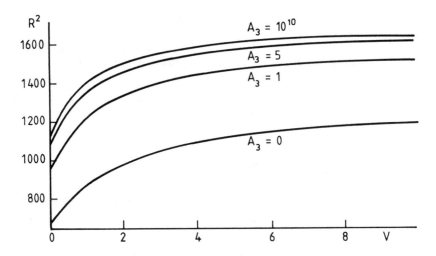

Figure 8.2. Energy stability curves; R^2 against surface viscosity, V.
(After McTaggart. Copyright 1984 by Pergamon Press PLC.
Reprinted with permission.)

For an increase in both V and A_3 the critical Rayleigh number of energy
theory increases. This agrees with Berg & Acrivos (1965) who found that
an increase in surface viscosity leads to an increase in stability. In their
work they cite the values $V = 1$ for a gaseous monolayer of stearic acid and
$V = 10^3$ for a condensed monolayer.

A_3	$V = 0$	$V = 1$	$V = 10^3$	$V \to \infty$
0	668.998	866.077	1294.838	1295.778
1	942.138	1206.687	1596.275	1596.970
5	1060.456	1345.452	1680.383	1680.976
∞	1100.650	1388.847	1707.199	1707.762

Table 8.1. Critical Rayleigh numbers of energy theory
for various surface viscosity, V, values
and thermal conductivity, A_3.
(After McTaggart. Copyright 1984 by Pergamon Press PLC.
Reprinted with permission.)

With $V = A_3 = 0$ her results agree with the value of 669 obtained by Davis & Homsy (1980) in their energy stability analysis for a non-deformable free surface with the Marangoni number set equal to zero. If $A_3 \to \infty$, $V = 0$, the boundary conditions reduce to the one rigid – one free conducting boundary conditions of the classical Bénard problem, and the value of 1100.65 coincides with the result obtained there. Similarly, when $V = A_3 \to \infty$ the boundary conditions correspond to the classical problem with conducting rigid boundaries and the value of 1708 obtained is thus in agreement.

To examine the effects of surface tension she considers $S \neq 0$. The critical Marangoni number is obtained from the relation $B = R_E^2 N$ with the results plotted in Figure 8.3.

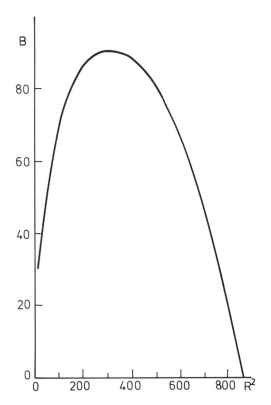

Figure 8.3. Energy stability curve; R^2 against the Marangoni number, B, with $V = 1, A_3 = 0, A_4 = 0.5$.
(After McTaggart. Copyright 1984 by Pergamon Press PLC. Reprinted with permission.)

In Figure 8.3, to the right of the maximum, there is agreement with the work of Davis (1969b) and Davis & Homsy (1980) in that increasing the

Marangoni number decreases the Rayleigh number.

We close this chapter with some remarks on the case where the linear relation (8.1) is not adequate. Cloot & Lebon (1986) use a sort of weakly nonlinear theory to address the convection problem when

$$\sigma = \sigma_m + \frac{1}{2}b(T - T_m)^2,$$

such a quadratic law, they point out, being appropriate to aqueous long chain alcohol solutions and some binary metallic alloys. A nonlinear energy analysis of this problem is interesting in so much as the method of Davis (1969b) does *not* appear to carry over. The problem arises because the perturbation surface stress has form

$$t^{ij}n_j = Mn^i(f\theta + \frac{1}{2}\theta^2) - Ma\, x^i_{;\alpha}a^{\alpha\beta}\theta_{;\beta} + Mx^i_{;\alpha}a^{\alpha\beta}\theta\theta_{;\beta}. \qquad (8.21)$$

Here M and Ma are Marangoni numbers, $x^i_{;\alpha}$ are derivatives of x^i with respect to the surface variables u^α, and $a^{\alpha\beta}$ is the surface metric tensor. The first term on the right of (8.21) does not play an important role in energy theory. The second term on the right corresponds to the surface temperature gradients in Davis' (1969b) analysis, but the $\theta\theta_{;\beta}$ term gives a problem. It gives rise to a *cubic* surface term in energy theory and evidently cannot be handled by the usual quadratic energy. One approach that will work is to add to the energy a piece of the L^4 integral of θ; this gives rise to a three-parameter eigenvalue problem like those described in chapter 4, where minimization is in one parameter while maximization is in *two* other parameters.

9

Convection in Generalized Fluids

9.1 The Bénard Problem for a Micropolar Fluid

If we allow the Newtonian constitutive equation for the stress to be replaced by something for a more exotic fluid then a variety of interesting convection problems arise. Nonlinear energy stability analyses for these fluids are only relatively recent. Slemrod (1978) developed energy theory for an integral-history linear viscoelastic convection problem, while Straughan (1983) and Franchi & Straughan (1988) developed analogous analyses for a fluid of second and of third grade, respectively; the third grade fluid equations are those of Fosdick & Rajagopal (1980).

For the various classes of polar fluids where the stress is not symmetric, some analyses are available. Nonlinear energy stability for a dipolar fluid heated from below is investigated in Straughan (1987). However, micropolar fluid theory has received more attention, and for the thermal problem presently appears to be richer; hence we now include some details of this.

The theory of micropolar fluids is due to Eringen (1964,1969,1972,1980) whose theory allows for the presence of particles in the fluid by additionally accounting for particle motion. In Eringen (1964) he introduced the theory for a simple microfluid and extended this in Eringen (1969) to allow deformation of local fluid elements (particles). This was further generalized in Eringen (1972) where he gave precise meaning to a thermomicropolar fluid, and in Eringen (1980) the theory appropriate to anisotropic fluids was elucidated.

The Bénard problem for the theory of Eringen (1972) was first examined by Datta & Sastry (1976) who studied stationary convection in the linear problem under an *assumption* on the sign of a thermal coupling term in the energy equation. Ahmadi (1976) used nonlinear energy stability theory on the same problem although he chose to neglect the coupling term, which gave rise to an interesting effect in Datta & Sastry (1976). Further use of the energy method was made by Lebon & Perez-Garcia (1981) who presented nonlinear stability results for the problem of Datta & Sastry (1976). The contribution of Payne & Straughan (1989) also looks at the effect of the thermal coupling term but concentrates primarily on the opposite sign to that selected by Datta & Sastry (1976) and Lebon & Perez-Garcia (1981). Payne & Straughan (1989) focus attention on the possibility of oscillatory convection (neglected in Datta & Sastry (1976)) and this yields a striking result; they also employ nonlinear energy stability but concentrate on the coefficient case not covered in Datta & Sastry (1976) and Lebon & Perez-Garcia (1981).

In Datta & Sastry (1976) it was shown that heating *both from above and below* could lead to stationary convection instabilities. The paper of Payne & Straughan (1989) provides an alternative route where stationary convection only occurs in the heated from below case. Furthermore, as the magnitude of the thermal coupling term is increased they predict a substantial decrease in the critical Rayleigh number (unlike Datta & Sastry (1976) and Lebon & Perez-Garcia (1981) where the opposite sign is chosen for the thermal coupling term). Chandra (1938) observed in his experiments that adding smoke particles to a layer of gas could lead to such a substantial decrease of the Rayleigh number at which convective motion commences. Since the particle spin associated with Eringen's (1972) theory could possibly be appropriate to the added dust situation described by Chandra (1938), there may be a justification for use of the analysis of Payne & Straughan (1989) for convection in a fluid fairly evenly interspersed with "particles", which may be dust, dirt, ice or raindrops, or other additives. Thus we believe that heuristically the results give reason to believe that the Eringen (1972) micropolar convection model may be applicable to geophysical or industrial convection contexts.

It is expedient to include the relevant equations of Eringen (1972) for an incompressible, isotropic thermomicropolar fluid. We take the microinertia moment tensor $j_{ik} = j\delta_{ik}$ where j is constant. The continuity, momentum, moment of momentum, and balance of energy equations are then (Eringen (1972), pp. 489, 490)

$$v_{i,i} = 0, \tag{9.1}$$

$$\rho\dot{v}_i = \rho f_i + t_{ki,k}, \tag{9.2}$$

$$\rho j\dot{\nu}_i = \rho\ell_i + \epsilon_{ikh}t_{kh} + m_{ki,k}, \tag{9.3}$$

$$\rho\dot{\epsilon} = t_{kh}b_{kh} + m_{kh}\nu_{h,k} + q_{k,k} + \rho h, \tag{9.4}$$

where $b_{kh} = v_{h,k} - \epsilon_{khr}\nu_r$, a superposed dot denotes the material derivative, $v_i, \nu_i, \epsilon(T)$ are fluid velocity, particle spin vector, and internal energy, T being temperature; $\rho, t_{ki}, f_i, \ell_i, m_{ki}, q_k$, and h are density (presumed constant except in the body force ρf_i term), stress tensor, body force, body couple, couple stress tensor, heat flux vector, and heat supply.

The constitutive equations are

$$t_{kh} = -\pi\delta_{kh} + \mu(v_{k,h} + v_{h,k}) + \bar{\kappa}(v_{h,k} - \epsilon_{kha}\nu_a), \tag{9.5}$$

$$m_{kh} = \bar{\alpha}\nu_{r,r}\delta_{kh} + \bar{\beta}\nu_{k,h} + \gamma\nu_{h,k} + \alpha\epsilon_{khm}T_{,m}, \tag{9.6}$$

$$q_k = \kappa T_{,k} + \beta\epsilon_{khm}\nu_{h,m}, \tag{9.7}$$

where π is the pressure and, in general, $\mu, \bar{\kappa}, \bar{\alpha}, \bar{\beta}, \gamma, \alpha, \kappa, \beta$ are functions of T. We treat $\mu, \bar{\kappa}, \bar{\alpha}, \bar{\beta}, \gamma, \alpha, \kappa$ as constant and in (9.1)–(9.4) ρ is assumed constant $(= \rho_0)$ except in the body force term in (9.2) where

$$\rho = \rho_0 \left[1 - A(T - T_0) \right], \tag{9.8}$$

A being the coefficient of thermal expansion of the fluid. It is further assumed that the body couple and heat supply are zero, i.e., $\ell_i \equiv h \equiv 0$.

The relevant equations are now obtained by inserting (9.5)–(9.8) into (9.1)–(9.4). However, some reduction of (9.4) is necessary to make the convection problem tractable. Equation (9.4) is

$$-\rho \frac{\partial \epsilon}{\partial T} \dot{T} + \left\{ (v_{h,k} - \epsilon_{khr} \nu_r) \left[\mu(v_{k,h} + v_{h,k}) + \bar{\kappa}(v_{h,k} - \epsilon_{kha} \nu_a) \right] \right.$$
$$\left. + \nu_{h,k} (\bar{\alpha} \nu_{r,r} \delta_{kh} + \bar{\beta} \nu_{k,h} + \gamma \nu_{h,k}) \right\} + \kappa \Delta T \tag{9.9}$$
$$+ \alpha \epsilon_{khm} T_{,m} \nu_{h,k} + \epsilon_{khm} \nu_{h,m} \beta_{,k} = 0.$$

Three approaches to reducing (9.9) have been advocated in the literature. All assume the quadratic terms (in {...} parentheses) are negligible. Such an approach is consistent with the analogous reduction for a Newtonian fluid.

Ahmadi (1976) further assumes the coefficient β is constant and also neglects the α term. The resulting Bénard problem is then of the form

$$u_t + Lu + N(u) = 0,$$

regarded as an equation in a Hilbert space, where the nonlinear term satisfies

$$< u, N(u) > \geq 0$$

in the appropriate inner product, and L is an unbounded, symmetric linear operator. Hence, no subcritical instabilities are possible, cf. chapter 4.

For a fluid contained in the layer $z \in (0, d)$ with gravity in the negative z-direction, the planes $z = 0, d$ kept at constant temperatures T_1, T_2 the steady solution of interest is

$$\bar{T} = -Bz + T_1, \qquad \bar{v}_i = 0, \qquad \bar{\nu}_i = 0, \tag{9.10}$$

where B is the temperature gradient.

The reduction of equation (9.9) by Datta & Sastry (1976) takes the coefficient β to be constant but retains the term $\alpha \epsilon_{khm} T_{,m} \nu_{h,k}$. This term gives rise to a non-zero contribution from $\partial \bar{T}/\partial z$, which makes the linear operator non-symmetric. The final approach by Lebon & Perez-Garcia (1981) neglects the α term, but they assume $\beta = \hat{\beta} T$ for $\hat{\beta}$ constant. This again leads to a contribution from $\partial \bar{T}/\partial z$ and hence some skew-symmetry. A richer structure is obtained by retaining a thermal coupling.

In Datta & Sastry (1976) and Lebon & Perez-Garcia (1981) the $\hat{\beta}$ term is assumed negative. However, thermodynamic arguments (Eringen (1972)) place only the following restrictions on the coefficients in (9.5)–(9.7):

$$2\mu + \bar{\kappa} \geq 0, \quad 2\mu + \kappa \geq 0, \quad \kappa \geq 0, \quad \bar{\kappa} \geq 0, \quad \mu \geq 0,$$

$$3\bar{\alpha} + \bar{\beta} + \gamma \geq 0, \quad \bar{\beta} + \gamma \geq 0, \quad \frac{2\kappa}{T}(\gamma - \bar{\beta}) \geq \left(\alpha - \frac{\beta}{T}\right)^2. \tag{9.11}$$

It would appear there is no reason why $\hat{\beta} = \beta T^{-1} < 0$. Indeed, the theoretical results of Payne & Straughan (1989) might suggest it preferable to require $\hat{\beta} > 0$.

The internal energy is given by $\epsilon = \psi + \eta T$ where $\eta = -\partial\psi/\partial T$. Thus,

$$\dot{\epsilon} = -\frac{\partial^2\psi}{\partial T^2}\,\dot{T}.$$

It is usual to assume $-\psi_{TT} = c$ (constant) and so we follow standard practice. Then, (9.9) may be reduced to

$$\rho_0 c\dot{T} = \kappa\Delta T + \hat{\beta}\epsilon_{khm}\nu_{h,m}T_{,k}. \tag{9.12}$$

To study the stability of (9.10) we define the perturbed quantities by

$$T = \bar{T} + \theta, \qquad v_i = \bar{v}_i + u_i, \qquad \nu_i = \bar{\nu}_i + \nu_i.$$

Important non-dimensional parameters are K, Γ, G, these being associated with the micropolar terms, $Ra = R^2$, the Rayleigh number, and the thermal term b, viz.,

$$K = \frac{\bar{\kappa}}{\mu}, \quad \Gamma = \frac{\gamma}{\mu d^2}, \quad G = \frac{\bar{\alpha} + \bar{\beta}}{\mu d^2}, \quad Ra = \frac{gABd^4 c\rho_0^2}{\kappa\mu}, \quad b = \frac{\hat{\beta}}{d^2 c\rho_0}.$$

The non-dimensionalized perturbation equations are

$$\begin{aligned}
u_{i,t} + u_k u_{i,k} &= R\theta\delta_{i3} - \pi_{,i} + \Delta u_i + K(\Delta u_i - \epsilon_{kir}\nu_{r,k}), \\
u_{i,i} &= 0, \\
Pr(\theta_{,t} + u_i\theta_{,i}) &= Rw + \Delta\theta - bR\xi + bPr\epsilon_{khm}\nu_{h,m}\theta_{,k}, \\
j(\nu_{i,t} + u_a\nu_{i,a}) &= K(\epsilon_{ikh}u_{h,k} - 2\nu_i) + G\nu_{a,ai} + \Gamma\Delta\nu_i,
\end{aligned} \tag{9.13}$$

where $\mathbf{u} \equiv (u, v, w)$ and $\xi = (\text{curl}\,\nu)_3$.

It is important to realize that the only term that makes the linearized version of (9.13) non-symmetric is the bR term in (9.13)$_3$.

Datta & Sastry (1976), Lebon & Perez-Garcia (1981), and Payne & Straughan (1989) consider equations (9.13) to hold in the layer $z \in (0, 1)$

with $\theta = \nu_i = 0$ on $z = 0, 1$ together with stress free boundary conditions on the surfaces $z = 0, 1$. We also assume a periodic structure in the x, y-plane.

9.2 Oscillatory Instability

A complete analysis of the linearized instability problem (with $b > 0$) is contained in Payne & Straughan (1989). We describe their findings for oscillatory convection only.

Equations (9.13) are linearized, and then a time dependence like

$$u_i(x_r, t) = e^{\sigma t} u_i(x_r)$$

is assumed with a similar form for ν_i, θ, p. The system of equations is reduced to a single equation in w, namely,

$$
\begin{aligned}
(\Gamma\Delta - 2K - j\sigma)&\Big\{(\Delta - Pr\sigma)\big[(1+K)\Delta^2 - \sigma\Delta\big] - R^2\Delta^*\Big\}w \\
&+ K\Big\{\Delta K(\Delta - Pr\sigma) - bR^2\Delta^*\Big\}\Delta w = 0,
\end{aligned}
\tag{9.14}
$$

where $\Delta^* \equiv \partial^2/\partial x^2 + \partial^2/\partial y^2$.

Assume a normal mode form

$$w = W(z)\Phi(x, y),$$

where $\Delta^*\Phi = -a^2\Phi$, and then with $D = d/dz$ the boundary conditions are

$$D^{(2n)}W = 0 \quad \text{at} \quad z = 0, 1, \forall n \geq 0.$$

Hence the solution to (9.14) is a half-range sine series of terms like $W = \sin n\pi z$. Denoting $\Lambda = n^2\pi^2 + a^2$, (9.14) reduces to

$$
\begin{aligned}
(\Gamma\Lambda + 2K + j\sigma)&\Big\{(\Lambda + Pr\sigma)\big[(1+K)\Lambda^2 + \sigma\Lambda\big] - R^2 a^2\Big\} \\
&- K\Big\{\Lambda K(\Lambda + Pr\sigma) + bR^2 a^2\Big\}\Lambda = 0.
\end{aligned}
\tag{9.15}
$$

To study oscillatory convection we select $\sigma = i\sigma_1$, $\sigma_1 \in \mathbf{R}$, in (9.15). By equating real and imaginary parts the following equations are obtained:

$$
\begin{aligned}
\sigma_1^2 = \frac{-1}{[j(1+Pr) + KPr(j-b)]} \\
\times \Big\{ \Lambda K^2 + K\Lambda^2 b(1+K) + \left(\frac{(Kb+\Gamma)\Lambda + 2K}{j}\right) \\
\times \Big(\Lambda\Gamma[1 + Pr(1+K)] + K^2 Pr + 2K(1+Pr)\Big)\Big\},
\end{aligned}
\tag{9.16}
$$

$$R_{\text{osc}}^2 = \frac{\Lambda^2}{a^2}(1+K)$$

$$+ \frac{1}{ja^2}\Big\{\Gamma\Lambda^3\big[1+Pr(1+K)\big]$$

$$+ 2K\Lambda^2(1+Pr) + K^2\Lambda^2 Pr\Big\}$$

$$+ \frac{\Lambda Pr}{a^2\big[j(1+Pr)+KPr(j-b)\big]} \tag{9.17}$$

$$\times \Big\{\Lambda K^2 + K\Lambda^2 b(1+K) + \left(\frac{(Kb+\Gamma)\Lambda+2K}{j}\right)$$

$$\times \Big(\Lambda\Gamma\big[1+Pr(1+K)\big]$$

$$+ K^2 Pr + 2K(1+Pr)\Big)\Big\}.$$

Since $\sigma_1 \in \mathbf{R}$, equation (9.16) yields much useful information. If $b < 0$ then we see immediately that overstable convection will not be possible unless b is large (in an appropriate sense): certainly it would have to be necessary that $b < -\Gamma/K$. For positive b, (9.17) immediately shows that a necessary condition for overstability is that

$$b > j\left(1 + \frac{1+Pr}{KPr}\right). \tag{9.18}$$

If b satisfies (9.18), then by grouping together Λ^3, Λ^2, and Λ terms in (9.17), we see that since j, K are likely small, R_{osc}^2 is *negatively* very large. A practical interpretation is that oscillatory convection is only possible when the layer is *heated from above*. Since the resulting critical Rayleigh numbers may well be enormous, indeed the numerical results of Payne & Straughan (1989) suggest $\sim 3.8 \times 10^6$, one could argue that such a situation would not be encountered in everyday life.

9.3 Nonlinear Energy Stability for Micropolar Convection

To study the nonlinear stability of (9.10) an L^2 energy, $E(t)$, is chosen, as

$$E = \frac{1}{2}(< u_i u_i > + j < \nu_i \nu_i >) + \frac{1}{2}\lambda Pr\|\theta\|^2, \tag{9.19}$$

where $\lambda\,(> 0)$ is a coupling parameter to be chosen. One could introduce another coupling parameter in front of the j term, but as there is symmetry between $(9.13)_1$ and $(9.13)_4$ the value of one for the second coupling parameter suffices. The functional (9.19) is also selected by Lebon & Perez-Garcia (1981).

We differentiate E and substitute from (9.13) to derive

$$\frac{dE}{dt} = RI - \mathcal{D}, \tag{9.20}$$

where the production term I and the dissipation $\mathcal{D}$ are in this case given by

$$I = (1 + \lambda) < \theta w > + b\lambda < \xi\theta >, \tag{9.21}$$

$$\begin{aligned}
\mathcal{D} = {} & < u_{i,j}u_{i,j} > + \lambda < \theta_{,i}\theta_{,i} > + K < \nu_i\nu_i > + \Gamma < \nu_{i,j}\nu_{i,j} > \\
& + K < (\epsilon_{ijk}u_{k,j} - \nu_i)(\epsilon_{irs}u_{s,r} - \nu_i) > + G\|\nu_{a,a}\|^2.
\end{aligned} \tag{9.22}$$

Define now

$$\frac{1}{R_E} = \max \frac{I}{\mathcal{D}}, \tag{9.23}$$

where the maximum is over the space of admissible functions. Then, from (9.20),

$$\frac{dE}{dt} \leq -\mathcal{D}\left(\frac{R_E - R}{R_E}\right),$$

and if $R < R_E$ then Poincaré's inequality together with strict forms of the thermodynamic inequalities (9.11) shows that

$$\frac{dE}{dt} \leq -kE,$$

for a positive constant k. Hence $R < R_E$ guarantees $E \to 0$ at least exponentially fast as $t \to \infty$, and we have nonlinear stability.

We put $\phi = \lambda^{1/2}\theta$ and calculate the Euler-Lagrange equations for R_E from (9.23), to find

$$\begin{aligned}
& (1 + K)\Delta u_i + K\epsilon_{ijk}\nu_{k,j} + R_E f\phi k_i = \pi_{,i}, \\
& u_{i,i} = 0, \qquad \Delta\phi + R_E\left(fw + \frac{1}{2}b\lambda^{1/2}\xi\right) = 0, \\
& \Gamma\Delta\nu_i + G\nu_{a,ai} + K(\epsilon_{ijk}u_{k,j} - 2\nu_i) - \frac{1}{2}\lambda^{1/2}b\epsilon_{3ji}R_E\phi_{,j} = 0,
\end{aligned} \tag{9.24}$$

where π is a Lagrange multiplier and

$$f = \frac{1 + \lambda}{2\lambda^{1/2}}.$$

Equations (9.24) are reduced to a single equation in w to find,

$$\begin{aligned}
& K\left(1 + \frac{1}{2}K\right)\Delta^3 w - \frac{1}{2}\Gamma(1 + K)\Delta^4 w \\
& + R_E^2\left[-Kf^2\Delta^* w - \frac{1}{8}\lambda b^2(1 + K)\Delta^2\Delta^* w \right. \\
& \left. + \frac{1}{2}(K\lambda^{1/2}bf + f^2\Gamma)\Delta\Delta^* w\right] = 0.
\end{aligned} \tag{9.25}$$

By employing a normal mode analysis (9.25) reduces to

$$R_E^2 = \frac{\Lambda^3}{a^2}\left(\frac{\Gamma(1+K)\Lambda + 2K + K^2}{2Kf^2 + \Lambda(K\lambda^{1/2}bf + f^2\Gamma) + \lambda b^2(1+K)\Lambda^2/4}\right). \qquad (9.26)$$

Then

$$Ra_E = \max_\lambda \min_{a^2} R_E^2(\lambda, a^2; \Gamma, K, b)$$

is found numerically.

Payne & Straughan (1989), like Datta & Sastry (1976) and Lebon & Perez-Garcia (1982), analyse the case of free-free boundaries.

Graphs of linear and energy critical Rayleigh numbers for varying b are presented in Figures 9.1, 9.2.

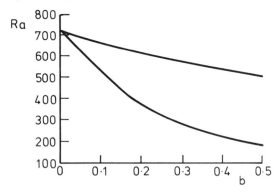

Figure 9.1. Ra against b. $\Gamma = 0.1$, $K = 0.1$. Upper curve represents linear results while lower one represents energy values.
(After Payne & Straughan. Copyright 1989 by Pergamon Press PLC. Reprinted with permission.)

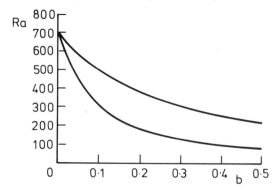

Figure 9.2. Ra against b. $\Gamma = 0.01$, $K = 0.1$. Upper curve represents linear results while lower one represents energy values.
(After Payne & Straughan. Copyright 1989 by Pergamon Press PLC. Reprinted with permission.)

The stationary convection results of Payne & Straughan (1989) show a drastic reduction in Ra for increasing b. Thus, an appropriate model could possibly account for the drastic reduction in Ra observed by Chandra (1938) when adding smoke particles to air. The energy results are useful in that they represent a *nonlinear unconditional* stability threshold and hence yield a region where *possible* subcritical instabilities may occur, although this region is quite large when $b \geq O(0.1)$. Perhaps by a more exotic choice of generalized energy one may be able to increase the nonlinear threshold substantially. The oscillatory convection results show that for Γ, K, j small, b needs to be relatively large for oscillatory convection to be possible. Even then, the numerical values of Ra are typically 10^3 larger than the ones observed for stationary convection in other parameter ranges, and this means that a very much larger *negative* temperature gradient would need to be imposed before the onset of such convection could be realized.

10
Time Dependent Basic States

10.1 Convection Problems
with Time Dependent Basic States

A physically important class of problem involves convection in a layer when the basic temperature is not simply a linear function of z, but where it depends also directly on time. For example, any attempt to grow semi-conductor crystals in space will be faced with the problem of a gravity field approximately 10^{-5} times as strong as that on Earth, and hence slight movement of the spacecraft will lead to a basic state that changes with time. Also, any geophysical convection problem driven by solar radiation comes into the category under consideration here since the Sun's heating effect follows either a diurnal cycle or a yearly one, depending on the timescale involved.

The first energy theory on convection problems involving time-dependent basic states would appear to be that of Homsy (1973) who made some general observations and obtained numerical results for fluid layers subject to rapid heating or cooling. Neitzel (1982) reexamines the numerical work pertinent to the analysis of Homsy (1973) and corrects some numerical results, while obtaining further results for other boundary conditions. Caltagirone (1980) treats the problem of a porous layer saturated with fluid and subject to instantaneous surface heating. His analysis contains a linearized study together with a comparison by the energy method, in addition to a check by a numerical finite difference solution in two-dimensions.

Homsy (1974) extends his energy theory to the convection problem where:

(a) the surface temperature varies sinusoidally with time, and

(b) gravity varies sinusoidally with time.

Energy theory for sinusoidally varying convection problems is also studied by Carmi (1974), and for the porous convection problem with time-periodic temperature boundary conditions by Chhuon & Caltagirone (1979). The latter paper also discusses a linear instability theory, experimental results, and the findings of a two-dimensional finite difference simulation. Before discussing Homsy's (1974) paper further we should point out that a linearized instability analysis of the gravity modulation problem

is given by Gresho & Sani (1970). Corresponding linearized analyses for
the surface temperature modulation problem are contained in Rosenblat
& Tanaka (1971) and in Yih & Li (1972). These papers utilize a normal
mode analysis in the horizontal x, y-directions, then assume an (Galerkin)
expansion of the z- part of the perturbation into a finite set of basis functions, thereby reducing the instability problem to one of studying a finite
set of coupled ordinary differential equations with time periodic coefficients.
Floquet theory (see e.g., Coddington & Levinson (1955)) is then used to
find a linear instability boundary. We do not describe the linearized analysis, although explicit details are given in the papers of Gresho & Sani
(1970), Rosenblat & Tanaka (1971), and Yih & Li (1972). Further work on
the time-dependent problem is contained in Finucane & Kelly (1976) and
Jhaveri & Homsy (1982) where further references may be found including
those on nonlinear work employing types of truncation techniques.

10.2 Energy Theory for Convection with Time-Varying Gravity and Surface Temperatures

The paper by Homsy (1974) is particularly lucid and does develop a very
useful theory for nonlinear energy stability for convection problems with
time dependent basic states, and so we describe the contents of this work.
Homsy (1974) considers an infinite layer of fluid bounded by surfaces $z =
0, d$, with surface temperatures

$$T = T_0 + T_s \cos \omega t, \quad z = 0,$$
$$T = T_1 + \delta T_s \cos \omega t, \quad z = d.$$

The basic temperature profile $\bar{\theta} = (T - T_1)/(T_0 - T_1)$ then satisfies the
diffusion equation

$$\frac{\partial \bar{\theta}}{\partial t} = \frac{\partial^2 \bar{\theta}}{\partial z^2} \tag{10.1}$$

and the boundary conditions

$$\bar{\theta} = 1 + a \cos \omega t, \quad z = 0,$$
$$\bar{\theta} = \delta a \cos \omega t, \quad z = 1, \tag{10.2}$$

where $a = T_s/(T_0 - T_1)$, ω in (10.2) being a dimensionless frequency.
    The solution $\bar{\theta}$ is found as

$$\bar{\theta} = 1 - z + a \operatorname{Re} \left\{ e^{i\omega t} \frac{\delta \sinh \beta z + \sinh [\beta(1 - z)]}{\sinh \beta} \right\}, \tag{10.3}$$

where

$$\beta = (1 + i)\sqrt{\frac{\omega}{2}}.$$

This solution is the one appropriate to surface modulation, but since gravity modulation effects appear directly in the momentum equation it is equally valid there too.

Perturbations to solution (10.3) satisfy

$$Pr^{-1}(u_{i,t} + u_j u_{i,j}) = -p_{,i} + \Delta u_i + R\theta k_i(1 + \epsilon \sin \omega t), \quad (10.4)$$

$$u_{i,i} = 0, \quad (10.5)$$

$$\theta_{,t} + u_i \theta_{,i} = \Delta\theta - Rw\frac{\partial\bar\theta}{\partial z}, \quad (10.6)$$

where θ is the temperature perturbation, and the $\epsilon \sin \omega t$ term represents the time-dependent variation in gravity field.

Homsy employs a coupling parameter λ and defines $\phi = \lambda^{1/2}\theta$ to obtain the energy equation

$$\frac{dE}{dt} = RI(t) - \mathcal{D}, \quad (10.7)$$

where

$$E = \frac{1}{2Pr}\|\mathbf{u}\|^2 + \frac{1}{2}\|\phi\|^2, \quad (10.8)$$

$$I(t) = \left\langle w\phi\left(\frac{1 + \epsilon \sin \omega t}{\lambda^{1/2}} - \lambda^{1/2}\frac{\partial\bar\theta}{\partial z}\right)\right\rangle, \quad (10.9)$$

$$\mathcal{D} = D(\mathbf{u}) + D(\phi). \quad (10.10)$$

From this point Homsy (1974) is able to proceed interestingly in two ways. The first is to follow Joseph's method (chapter 4). Hence, define

$$\frac{1}{\rho_\lambda} = \max_{\mathcal{H}} \frac{I}{\mathcal{D}}. \quad (10.11)$$

The function $\rho_\lambda(t)$ is periodic in t with period $2\pi/\omega$. Define $R_{s,\lambda}$ by

$$R_{s,\lambda} = \min_{t \in [0, 2\pi/\omega)} \rho_\lambda(t).$$

Then, the usual analysis shows $R < R_{s,\lambda}$ guarantees nonlinear stability. Finally, optimize in λ to find a nonlinear stability threshold R_s given by

$$R_s = \max_\lambda \min_{t \in [0, 2\pi/\omega)} \rho_\lambda(t). \quad (10.12)$$

Homsy (1974) then argues that for time-periodic problems the condition $R < R_s$, while guaranteeing all disturbances decay rapidly, may be overly conservative. He proposes an alternative based on an idea of Davis & von

Kerczek (1973), which allows the amplitude of a disturbance to possibly *increase* over a cycle, provided the disturbance decays to zero over many cycles, i.e., as $t \to \infty$.

Basically the argument of Homsy (1974) is to define

$$\nu_\lambda(t) = \max_{\mathcal{H}} \left(\frac{RI - \mathcal{D}}{E} \right), \tag{10.13}$$

and note (provided the maximum in (10.13) exists) that (10.7) yields

$$\frac{dE}{dt} \le \nu_\lambda(t)E.$$

This integrates to

$$E(t) \le E(0) \exp \int_0^t \nu_\lambda(s)ds.$$

The basic state is then stable provided

$$\int_0^{2\pi/\omega} \nu_\lambda(s)\, ds < 0. \tag{10.14}$$

The idea is to choose λ such that (10.14) holds for as large a number R as possible: this number is denoted by R_A. Thus the criterion for asymptotic stability is $R < R_A$.

Homsy (1974) treats gravity variation and surface temperature variation separately, obtaining interesting numerical results for both cases. We report only the former. For this case choose $a = 0$ and then observe that $\partial\bar\theta/\partial z = -1$. The strong (decay for all disturbances) limit R_s arises from

$$\frac{1}{\rho_\lambda} = \max_{\mathcal{H}} \frac{<w\phi>}{\mathcal{D}} \Lambda(t),$$

where

$$\Lambda(t) = \frac{1 + \lambda + \epsilon \sin \omega t}{\sqrt{\lambda}}.$$

The Euler-Lagrange equations for this are

$$u_{i,i} = 0, \qquad \Delta u_i + \frac{1}{2}\rho_\lambda\phi k_i\Lambda(t) = \pi_{,i}\,,$$

$$\Delta\phi + \frac{1}{2}\rho_\lambda\Lambda(t)w = 0. \tag{10.15}$$

Homsy observes that these are the same as for the steady Bénard problem if R in that case is replaced by $\frac{1}{2}\rho_\lambda\Lambda(t)$. Hence, he concludes

$$\frac{1}{2}\rho_\lambda\Lambda(t) = R_L\,,$$

where $R_L^2 \approx 1708$ for rigid boundaries. His global limit is found from

$$R_s = \max_\lambda \min_t \rho_\lambda(t) = \max_\lambda \frac{2R_L \lambda^{1/2}}{1 + \epsilon + \lambda} = \frac{R_L}{(1 + \epsilon)^{1/2}}.$$

Hence, if the Rayleigh number $Ra\,(= R^2)$ satisfies $Ra < R_L^2/(1 + \epsilon)$ the layer is strongly stable in that disturbances decay rapidly, monotonically.

To improve this result, however, requires use of the result (10.14). The maximum problem for this is (10.13) which leads to the different Euler-Lagrange equations,

$$\begin{aligned}
\nu_\lambda \frac{u_i}{Pr} &= -\pi_{,i} + \Delta u_i + \frac{1}{2} R\Lambda(t)\phi k_i, \\
\nu_\lambda \phi &= \Delta\phi + \frac{1}{2} R\Lambda(t)w, \quad u_{i,i} = 0.
\end{aligned} \tag{10.16}$$

Homsy (1974) reports he was unable to solve (10.16) in general but instead he made significant partial progress with the free-free boundary case as follows. The curlcurl of $(10.16)_2$ is taken and then the third component of this is

$$\nu_\lambda Pr^{-1}\Delta w = \Delta^2 w + \frac{1}{2} R\Lambda(t)\Delta^* \phi. \tag{10.17}$$

The system $(10.16)_3, (10.17)$ is solved by taking

$$\begin{aligned}
w &= \hat{w} f(x, y) \sin \pi z, \\
\phi &= \hat{\phi} f(x, y) \sin \pi z,
\end{aligned}$$

where

$$\Delta^* f + a^2 f = 0.$$

For non-trivial $\hat{w}, \hat{\phi}$ he reduces $(10.16)_3, (10.17)$ to

$$\nu_\lambda = \frac{1}{2} h \left\{ -(1 + Pr) + \left[(1 - Pr)^2 + Pr\left(\frac{Ra}{R_L^2} \right) \Lambda^2 \right]^{1/2} \right\}, \tag{10.18}$$

where

$$h = \pi^2 + a^2, \qquad R_L^2 = \frac{h^3}{\alpha^2}.$$

He again reports that it was not possible to proceed and apply condition (10.14). Instead, he investigates the cases $Pr = 1$, $Pr \to 0$, and $Pr \to \infty$. For $Pr = 1$, he finds

$$\frac{\nu_\lambda}{h} = -1 + \frac{R}{R_L}\left(\frac{1 + \epsilon \sin \omega t + \lambda}{2\sqrt{\lambda}} \right).$$

Condition (10.14) is then satisfied whenever

$$R < \frac{2\lambda^{1/2}}{1 + \lambda} R_L.$$

By selecting $\lambda = 1$ he then finds $R_A = R_L$; this is undoubtedly due to the fact that for $Pr = 1$, (10.16) is a symmetric linear system.

For $Pr \to \infty$ Homsy (1974) finds

$$\nu_\lambda = \frac{\alpha^2}{h^2} \left(\frac{Ra\Lambda^2}{4} - R_L^2 \right),$$

condition (10.14) yields

$$R^2 \leq \frac{4\lambda R_L^2}{(1 + \lambda)^2 + \frac{1}{2}\epsilon^2},$$

and $\lambda^2 = 1 + \frac{1}{2}\epsilon^2$ maximizes, giving the asymptotic stability limit

$$R_A = \frac{2R_L}{1 + (1 + \frac{1}{2}\epsilon^2)^{1/2}}, \qquad Pr \to \infty. \qquad (10.19)$$

By a similar argument he derives the same R_A bound as (10.19), for $Pr \to 0$.

Homsy (1974) observes that $R_A(Pr) = R_A(Pr^{-1})$ and $R_s \leq R_A \leq R_L$. Homsy's paper is an important one and deserves full recognition in the nonlinear energy stability literature.

Another useful contribution to energy theory in time-dependent convection is by Gumerman & Homsy (1975) who treat the problem of a layer of fluid impulsively cooled at its free surface, with surface tension effects taken into account. This thereby combines the analysis of Homsy (1973) and that of Davis (1969b), the surface tension theory described in chapter 8. In Gumerman & Homsy (1975) they also obtain estimates on the *onset time* for the instability to commence. The physical motivation for the paper of Gumerman & Homsy (1975) is to model evaporating liquids subject to convective instabilities induced by surface tension. In order for the surface effects to dominate the buoyancy effect, they restrict attention to thin layers, 3 mm or less. The evaporation is accounted for by a condition of constant heat flux from the upper surface. The lower surface condition is an isothermal one, thereby allowing a comparison of stability results for large times with those of Davis (1969b), see chapter 8. The essential difference with that of chapter 8 is that a quiescent layer of fluid, depth d, rests on a horizontal, constant temperature plate with a free upper surface. At time $t = 0$, the layer is impulsively cooled by a constant outwardly directed heat flux and this results in a *basic* (dimensionless) temperature profile of form

$$\bar{T}(z, t) = T_0 - z + 2 \sum_{m=0}^{\infty} \frac{(-1)^m}{M^2} \exp\left(-M^2 t\right) \sin\left(Mz\right), \qquad (10.20)$$

where

$$M(m) = \frac{(2m + 1)\pi}{2}$$

and T_0 is the initial (dimensionless) temperature of the layer. The nonlinear energy stability study is then akin to that at the beginning of chapter 8, with boundary conditions (8.2), *but with the basic temperature given by* (10.20). Of course, this changes the analysis considerably since the production term in the energy equation before (8.3) now involves a time-dependent basic state. The analysis again employs Homsy's time dependent version of energy theory and the Davis-von Kerczek theory. By comparing their theoretical results with experimental measurements for propyl alcohol, methyl alcohol and acetone, they conclude that the experimental results are in broad agreement with the theoretical ones.

Up to this point the book has concentrated almost exclusively on applications of energy methods to nonlinear stability problems in convection. For the remainder of the book we change direction somewhat and discuss four relatively new and technologically important theories involving convection phenomena. Of necessity, we largely restrict attention to linearized instability analyses after presentation of the necessary equations to describe the theory, since most available stability work has been performed within the confines of the linear theory. The few energy results I am aware of, however, are discussed.

11
Electrohydrodynamic and Magnetohydrodynamic Convection

11.1 Comments on the MHD Bénard Problem and a Brief Review of Thermo-Convective Electrohydrodynamic Instability

As Rosensweig (1985) points out, the interaction of electromagnetic fields and fluids has been attracting increasing attention due to applications in many diverse areas. He writes that the subject may be divided into three main categories:

1. *Electrohydrodynamics* (EHD), the branch of fluid mechanics concerned with electric force effects;

2. *Magnetohydrodynamics* (MHD), the study of the interaction between magnetic fields and fluid conductors of electricity; and

3. *Ferrohydrodynamics* (FHD), which deals with the mechanics of fluid motion induced by strong forces of magnetic polarization.

Topics 1 and 3 are relatively new and are attracting increasing attention in the theoretical and engineering literature. Rosensweig (1985) pp. 1,2, explains succinctly the differences between the above topics as follows.* *In MHD the body force acting on the fluid is the Lorentz force that arises when electric current flows at an angle to the direction of an impressed magnetic field. However, in FHD there need be no electric current flowing in the fluid, and usually there is none. The body force in FHD is due to polarisation force, which in turn requires material magnetisation in the presence of magnetic field gradients or discontinuities. Likewise, the force interaction arising in EHD is often due to free electric charge acted upon by an electric force field. In comparison, in FHD free electric charge is normally absent, and the analog of electric charge, the monopole, has not been found in nature. An analogy between EHD and FHD arises, however, for charge-free electrically polarizable fluids exposed to a gradient electric field. A major difference from FHD is the magnitude of the effect, which is normally much smaller in the electrically polarizable media.*

---

* Used by permission of Cambridge University Press.

In this book we concentrate on convection-like problems in EHD and FHD, in chapters 11 and 12, respectively. Most of the analysis described is via linear instability theory since energy theory has been so far successful only in certain cases. Indeed, this is a rich area for future research. Before doing this, however, a few remarks are in order on the important MHD convection problem.

The magnetohydrodynamic convection problem is a very important one in so much that it has intrinsic applications to the behaviour of planetary and stellar interiors, and in particular, to the behaviour inside of the Earth. Consequently, there has been much work on this problem, much of it by weakly nonlinear analysis, but also very useful bounds have been obtained by using a technique not unrelated to the energy method, the variational theory of turbulence. This is critically reviewed by Fearn et al. (1988), see also Busse (1988), and for this and other related dynamo topics see Roberts (1987a,b,1988a,b). I am unaware of any direct use of energy theory on the convection dynamo problem, apart from the Cowling anti-dynamo type theorems, but it is an area where I believe it could be usefully developed, especially in one of the limit theories where the full problem is reduced to something more tractable.

The first use of the energy method in magnetohydrodynamics, along the lines reviewed here was by Rionero (1967,1968,1971). His contributions are fundamental, being the first to establish existence of a maximising solution in the energy variational problem. In Rionero (1971) he also included the Hall effect. Linear theory, Chandrasekhar (1981), shows that as the field strength is increased for the MHD convection problem with the magnetic field perpendicular to the layer, the magnetic field has a strongly inhibiting effect on the onset of convective motion. The first analyses to confirm this stabilizing effect from a nonlinear energy point of view are due to Galdi & Straughan (1985a) and Galdi (1985). Galdi (1985) introduced a highly non-trivial generalized energy that contains gradients as well as the fields themselves; this energy having some resemblance to the one needed to obtain stabilization in the rotating Bénard problem, see chapter 6.

It would appear that studies of convection-like instabilities in insulating fluid layers subject to temperature gradients and electrical potential differences across the layer, first gained impetus in the 1960s. It is not the object of this book to present a critical review of the early literature on this topic. Instead we present findings of work most relevant to this book, commencing with that of Roberts (1969) and Turnbull (1968a,b). Roberts (1969) reports on experiments of Gross (1967) in which a layer of insulating oil is confined between horizontal conducting planes and is heated from above and cooled from below. Despite the fact that this situation is gravitationally stable from a thermal point of view, Gross observed that when a vertical electric field of sufficient strength is applied across the layer, a tesselated pattern of motions is observed, in a manner similar to that of standard Bénard convection. Gross suggested that this phenomenon may be due to variation of the dielectric constant, ϵ, of the fluid with temperature.

Roberts (1969) performed a theoretical analysis to investigate Gross' suggestion and also investigated an alternative model that allows for the convection to be due to free charge conducted in the layer.

11.2 The Investigations of Roberts and of Turnbull

Roberts (1969) first allows the dielectric constant of the fluid to vary with temperature. He considers a homogeneous insulating fluid at rest in a layer with vertical, parallel applied gradients of temperature, T, and electrostatic potential, V. He assumes the layer depth is d, and denotes by $\mathbf{D}$ the electric displacement. The body force, per unit volume, on an isotropic dielectric fluid is then given by (Landau et al. (1984), eq. (15.15))

$$\mathbf{f} = -\operatorname{grad} p + \frac{\rho}{8\pi} \operatorname{grad} \left\{ E^2 \left(\frac{\partial \epsilon}{\partial \rho} \right)_T \right\} - \frac{E^2}{8\pi} \left(\frac{\partial \epsilon}{\partial T} \right)_\rho \operatorname{grad} T, \qquad (11.1)$$

where $p, \mathbf{E}$, and ρ denote pressure, electrical field, and density. The appropriate Maxwell equations are

$$\operatorname{div} \mathbf{D} = 0, \qquad \operatorname{curl} \mathbf{E} = 0, \qquad (11.2)$$

with $\mathbf{D}$ given for an isotropic material by

$$\mathbf{D} = \epsilon \mathbf{E}. \qquad (11.3)$$

Since the curl of $\mathbf{E}$ vanishes an electric potential V exists such that

$$\mathbf{E} = -\operatorname{grad} V.$$

Roberts (1969) examines a constant density, ρ_0, Newtonian fluid (apart from the thermal buoyancy term) for which the momentum and continuity equations are

$$u_{i,t} + u_j u_{i,j} = g_i + \nu \Delta u_i, \qquad (11.4)$$
$$u_{i,i} = 0, \qquad (11.5)$$

where $\mathbf{u}$ is velocity, ν viscosity, and for a linear thermal body force,

$$g_i = -\omega_{,i} - \frac{E^2}{8\pi\rho_0} \left(\frac{\partial \epsilon}{\partial T} \right)_\rho T_{,i} - g[1 - \alpha(T - T_0)]k_i, \qquad (11.6)$$

g being gravity, α thermal expansion coefficient, and $\mathbf{k} = (0,0,1)$, with

$$\omega = \frac{p}{\rho_0} - \frac{E^2}{8\pi} \left(\frac{\partial \epsilon}{\partial \rho} \right)_T .$$

The energy equation governing the temperature field is chosen as

$$T_{,t} + u_i T_{,i} = \kappa \Delta T, \tag{11.7}$$

κ being thermal diffusivity.

The geometry of Roberts (1969) consists of the fluid occupying the region between the planes $z = \pm \frac{1}{2} d$, which are maintained at uniform, but different temperatures $T = \pm \frac{1}{2} \beta d$, for β constant. A uniform electric field is applied in the z–direction. The equilibrium solution is denoted by an overbar and then

$$\bar{\mathbf{u}} = \mathbf{0}, \qquad \bar{T} = \beta z, \qquad \bar{\epsilon} \bar{E}_3 = \epsilon_m E_m \text{ (constant)}, \tag{11.8}$$

or

$$\bar{E} \equiv \bar{E}_3 = -\frac{d\bar{V}}{dz} = \frac{\epsilon_m E_m}{\epsilon(\bar{T})} = \frac{\epsilon_m E_m}{\epsilon(\beta z)},$$

with $\bar{\omega}$ being determined from (11.4). Denoting now the *perturbations* to $\bar{V}, \bar{\mathbf{E}}, \bar{\epsilon}, \bar{\omega},$ and $\bar{T}$ by $V', \mathbf{E}', \epsilon', \omega,$ and θ, the *linearized* equations for instability are found to be

$$\epsilon' = \left(\frac{\partial \epsilon}{\partial T} \right)_{\rho_0} \theta,$$

$$\Delta V' = \bar{E} \frac{\partial}{\partial z} \left(\frac{\epsilon'}{\bar{\epsilon}} \right) - \frac{1}{\bar{\epsilon}} \frac{d\bar{\epsilon}}{dz} \frac{\partial V'}{\partial z},$$

$$u_{i,t} = -\pi_{,i} + \nu \Delta u_i - B \delta_{i3} + \alpha g \delta_{i3} \theta,$$

$$\theta_{,t} = -\beta w + \kappa \Delta \theta,$$

where

$$w = u_3, \qquad \pi = \omega + \frac{\epsilon' \bar{E}^2}{8\pi},$$

$$B = \frac{\bar{E}}{4\pi \rho_0} \frac{d\bar{\epsilon}}{dz} \left(\frac{\epsilon' \bar{E}}{\bar{\epsilon}} - V'_{,z} \right).$$

At this point Roberts (1969) assumes

$$\epsilon = \epsilon_m [1 + \eta T], \tag{11.9}$$

η constant, with ϵ_m being the value of the dielectric constant at a reference temperature $T_m = 0$ deg. His Boussinesq approximation discards any term involving ηT when a similar term occurs not containing that factor. (This is a subtle point when considering a nonlinear analysis and has important mathematical ramifications connected to elliptic estimates, see Galdi & Straughan (1990).)

The linearized equation for θ is then derived by Roberts (1969) to be

$$\left(\frac{\partial}{\partial t} - \Delta \right) \left(Pr \frac{\partial}{\partial t} - \Delta \right) \Delta^2 \theta = L \Delta^{*2} \theta + Ra \Delta \Delta^* \theta, \tag{11.10}$$

where $Pr = \nu/\kappa$ is the Prandtl number and

$$Ra = \frac{\alpha\beta d^4 g}{\kappa\nu}, \qquad L = \frac{\epsilon_m E_m^2}{4\pi\rho_0}\frac{\beta^2\eta^2 d^4}{\kappa\nu},$$

are, respectively, the Rayleigh number and a parameter effectively measuring the potential difference between the planes. Upon seeking a normal mode representation $\theta = F(z)e^{i(lx+my)+st}$, with $D = d/dz$ and $a^2 = l^2 + m^2$ denoting the square of the wavenumber, equation (11.10) becomes

$$(D^2 - a^2 - s)(D^2 - a^2 - Pr\,s)(D^2 - a^2)^2 F = La^4 F - Ra\,a^2(D^2 - a^2)F. \quad (11.11)$$

For fixed surfaces the boundary conditions are

$$W = DW = 0, \qquad F = 0, \tag{11.12}$$

where $W(z)$ is the z–part of w in its normal mode representation. To interpret these as conditions on F we observe that the electric field continues outside the layer $z \in (-\frac{1}{2}, \frac{1}{2})$ and so the potential satisfies Laplace's equation there. For example, in $z > \frac{1}{2}$, if $\hat{V}'$ denotes the electric field perturbation, $\hat{V}' = G(z)\exp[i(lx + my) + st]$, then

$$(D^2 - a^2)G = 0,$$

and for $G \to 0$ as $z \to \infty$, necessarily

$$G = Ae^{-az}, \tag{11.13}$$

for some constant A. If $k = \epsilon_m/\hat{\epsilon}$, $\hat{\epsilon}$ being the dielectric constant of the solid in $z > \frac{1}{2}$, then on $z = \frac{1}{2}$,

$$\hat{V}' = V' \quad \text{and} \quad \frac{\partial\hat{V}'}{\partial z} = k\frac{\partial V'}{\partial z}.$$

From (11.13) this leads to

$$aG + kDG = 0, \quad \text{on} \quad z = \frac{1}{2}. \tag{11.14}$$

For $z < \frac{1}{2}$ composed of the same material,

$$aG - kDG = 0, \quad \text{on} \quad z = -\frac{1}{2}. \tag{11.15}$$

Roberts then shows conditions (11.12), (11.14), and (11.15) convert to the following conditions on F, on the planes $z = \pm 1/2$,

$$F = D^2 F = D(D^2 - Pr\,s)F = 0,$$
$$[D^2(D^2 - s)(D^2 - Pr\,s) + Ra\,a^2](DF \pm kaF) = 0, \tag{11.16}$$

where $\mathcal{D}^2 = D^2 - a^2$. Thus, the eighth-order equation (11.11) is to be solved subject to the eight boundary conditions (11.16). Roberts (1969) investigates only stationary convection and finds

$$\min_{a^2} Ra(a^2) \qquad \text{for fixed } L,$$

numerically.

The results are given below.

Ra	L	a
-1000	3370.077	3.2945
-500	2749.868	3.2598
0	2128.696	3.2260
500	1506.573	3.1929
1000	883.517	3.1606
1707.762	0	3.1162

Table 11.1. Critical linear Rayleigh numbers
against the parameter L and wavenumber a.
(After Roberts. Copyright 1969 by Oxford University Press.
Reprinted with permission.)

Of course in this table it must be remembered that L contains β^2 and so care must be exercised with a direct comparison of critical Ra against voltage difference, the quantities measured experimentally be Turnbull (1968b).

Roberts (1969) concludes his investigation of the above model by including physical values for the various parameters that arise. For a temperature difference of 50°C across a gap of 1mm he notes that Gross (1967) observes a tesselated pattern for what corresponds to a value of $L \approx 2 \times 10^{-4}$. According to his theory Ra_{crit} should be near 1708 whereas Gross deliberately enforced Ra to be negative. Roberts concluded that the instability mechanism of Gross' experiments was not that of the foregoing model.

The second model studied by Roberts (1969) was also essentially the one investigated independently by Turnbull (1968a), and is the one that has effectively been used since, albeit with modification. This model assumes the variation in the dielectric constant is not important but the fluid is weakly conducting and the conductivity varies with temperature. Turnbull (1968b) is motivated by his experimental results, indeed on p. 2601 his graphs of conductivity, σ, against temperature indicate that for corn oil the conductivity increases about $7\frac{1}{2}$ times in a temperature change from 20°C to 50°C. For his other working fluid, castor oil, the change over the same temperature interval would appear to be four-fold. Turnbull (1968a)

p. 2588 remarks, *In liquids the density typically varies about 0.1% per °C,
and thus, the dielectric constant varies about the same amount. However,
the electrical conductivity has a much larger variation with temperature.
For example, the conductivity of both corn oil and castor oil varies about
5% per °C.*

While Roberts' (1969) second model and that of Turnbull (1968a) are
essentially the same, Turnbull allows a linear variation of viscosity with
temperature and a quadratic variation of electrical conductivity. Mathematically, however, there is a difference in that Roberts concentrates on
stationary convection whereas Turnbull (1968a) does allow for oscillatory
convection although he does not solve the eigenvalue problem correctly. He
instead argues (p. 2592) that a Fourier sine series solution should be adequate since, ...*Ohm's law is only an approximation for poorly conducting
liquids, and, therefore, it makes no sense to solve the equations exactly
since the model is only an approximation.*

We include below graphs of the *experimental* results of Turnbull (1968b)
on the onset of convection. While those for corn oil are perhaps not too
convincing that increasing the upper temperature allows a smaller voltage
difference to trigger instability, there is certainly more evidence of this fact
for castor oil.

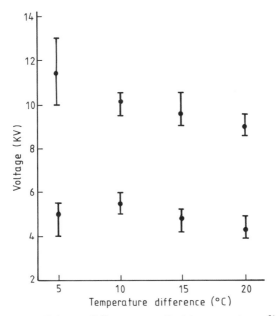

Figure 11.1. Voltage difference against temperature difference,
for instability in corn oil. Bottom of tank maintained at 25 °C.
Upper points together with error bounds are for 2 inch high tank,
lower are for 1 inch high tank.
(After Turnbull. Copyright 1968 by American Institute of Physics.
Reprinted with permission from author and publisher.)

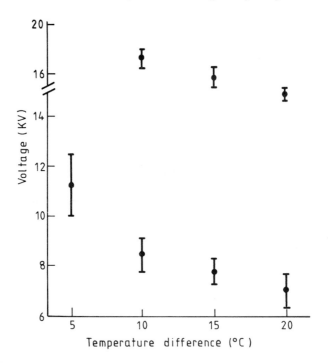

Figure 11.2. Voltage difference against temperature difference, for instability in castor oil. Bottom of tank maintained at 20 °C. Upper points together with error bounds are for 2 inch high tank, lower are for 1 inch high tank.
(After Turnbull. Copyright 1968 by American Institute of Physics. Reprinted with permission from author and publisher.)

For completeness we now include a brief description of Roberts' analysis and results.

Ohm's law and the equation of charge conservation are

$$\mathbf{j} = \sigma\mathbf{E} + Q\mathbf{u}, \tag{11.17}$$

$$\frac{\partial Q}{\partial t} + \operatorname{div}\mathbf{j} = 0, \tag{11.18}$$

where $\mathbf{j}$ is current and Q the volume space charge.

Maxwell's equations give

$$4\pi Q = \operatorname{div}\mathbf{D} = \epsilon\operatorname{div}\mathbf{E} = -\epsilon\Delta V.$$

The body force term in the momentum equation is now

$$\mathbf{g} = -\operatorname{grad}\frac{p}{\rho} + \frac{Q\mathbf{E}}{\rho},$$

together with a thermal buoyancy term, if required.

The equilibrium state has

$$\bar{E} = -\frac{d\bar{V}}{dz} = \frac{\sigma_m E_m}{\sigma(\beta z)}.$$

With a linear conductivity law of form

$$\sigma = \sigma_m\{1 + \bar{\eta}(T - T_m)\}$$

Roberts (1969) develops a Boussinesq approximation and seeking instability due to stationary convection he arrives at the eigenvalue equation

$$(D^2 - a^2)^3 F = -Ra\, a^2 F - Ma^2 DF, \qquad (11.19)$$

where M is a dimensionless measure of the variation of electrical conductivity with temperature, defined by

$$M = \frac{\epsilon E_m^2}{4\pi\rho_0} \frac{\beta\bar{\eta}d^3}{\nu\kappa}.$$

For rigid boundaries at constant temperatures the boundary conditions for (11.19) are

$$F = D^2 F = D(D^2 - a^2)F = 0 \qquad \text{on} \qquad z = \pm\frac{1}{2}. \qquad (11.20)$$

Roberts (1969) solves (11.19), (11.20) numerically, seeking the minimum over a^2. He notes that numerically Ra_{crit} is seen to be an even function of M. His results are given in Table 11.2.

Ra	M	a
1707.762	0	3.116
1722.826	200	3.112
1767.670	400	3.068
1841.106	600	3.008
1941.160	800	2.926
2065.034	1000	2.824

Table 11.2. Critical values of the Rayleigh number against the wavenumber a and the parameter M. (After Roberts. Copyright 1969 by Oxford University Press. Reprinted with permission.)

11.3 Further Work on Thermo-Convective EHD Instability

Subsequent work would appear to have followed the second approach described above, focusing on conductivity variation with temperature and ramifications of equations (11.17), (11.18). In particular, Takashima & Aldridge (1976) take ϵ linear in T, σ quadratic in T and study stationary convection. Bradley (1978) treats a linear conductivity but resolves the discrepancy between Roberts (1969) stationary convection and Turnbull's (1968a) oscillatory convection results by delimiting parameter ranges where overstable convection will be preferred, taking into account the correct boundary conditions.

Another connected approach had been followed by P. Atten and his coworkers from the late 1960s. References to this body of work are given by Worraker & Richardson (1979), who point out these writers concentrated on mobility models of charge transport. They write (pp. 29–30),* *...electrical conduction in well-filtered liquid hydrocarbons (e.g. n-hexane, transformer oil, kerosene) is governed by the presence of neutral covalent electroactive impurities that undergo oxidation and reduction reactions at the electrodes. The charged impurities are then assumed to drift down an electrical potential gradient with a velocity that is linearly proportional to the local field strength, the constant of proportionality being called the carrier or ionic mobility.*

This mobility approach allows us to consider the practical possibility of controlling the species of impurity, the intensity of injection and the position of the emitting electrode independently of any applied temperature gradient. Furthermore, there is the possibility of a charge-induced fluid dynamic instability even in an isothermal system. A conductivity model, however, does not permit consideration of such details nor does it admit the possibility of an isothermal instability.

In referring to the work of Turnbull (1968a), Roberts (1969), Takashima & Aldridge (1976), and Bradley (1978), Worraker & Richardson (1979, p. 30) write, ... *their results still leave doubt as to the underlying mechanisms responsible for the observed fluid motions. Since we believe that impurities will have a profound effect on the behaviour of such an experimental system, we have chosen to investigate a model that is consistent with much better controlled electrochemical experiments.*

The equations of Worraker & Richardson (1979) are still (11.4)–(11.7) and (11.18). However, their constitutive theory takes

$$\mathbf{j} = Q\mathbf{u} + QK\mathbf{E}, \qquad (11.21)$$

where K is a "charge carrier mobility," which they assume depends on

---

* Used by permission of Cambridge University Press

temperature,
$$K = K_0\left[1 + k_1(T - T_0)\right]. \tag{11.22}$$
They also allow a linear variation of ϵ with temperature. Their linear analysis, which concentrates on stationary convection, is carefully carried out and they conclude (p. 43), ... *Perhaps the most striking feature of the above analysis is that it demonstrates the importance of the sign of the temperature gradient in relation to the emitting electrode. It suggests that a system with an emitter cooler than the collector is more susceptible to stationary instability than one with the opposite temperature gradient.* Castellanos et al. (1984a,b) continue in this vein analysing a variety of effects with temperature dependent mobility models, investigating both stationary and oscillatory convection.

Martin & Richardson (1984) return to the temperature dependent conductivity model taking an Arrhenius exponential dependence on temperature in their linear stability analysis. They conclude that a convective instability cannot be predicted by a simple conductivity model but the true picture is probably one with the instability driven by charge injection (mobility model), but modified by residual conduction.

Rodriguez et al. (1986) further continued the above studies allowing charge injection strengths of any finite magnitude. They draw specific attention to the fact that (p. 2115), ... *it is possible to consider two well differentiated mechanisms of charge generation: (i) that resulting in the bulk of the liquid and induced by conductivity gradients and (ii) the injection of charge from one (unipolar) or both (bipolar) electrodes.*

The linear instability picture is now fairly complete for the conductivity and mobility models, covered comprehensively in the above cited works and references therein. Nonlinear studies, on the other hand, seem in relative infancy. Richardson (1988) studies nonlinear bifurcation on the basis of a simplified model of two coupled ordinary differential equations. The nonlinear aspect certainly would appear well worth further consideration. Applications of energy theory in this field are almost non-existent, and this should certainly be worth pursuing.

11.4 Charge Injection Induced Instability and the Nonlinear Energy Stability Analysis of Deo and Richardson

We describe a very interesting contribution by Deo & Richardson (1983) who apply a generalized energy analysis to the problem of the title of this section. In connection with the work of the previous sections it studies a mobility type model, but under isothermal conditions. However, in the context of this book it is important in that it is the first study of energy stability in convective electrohydrodynamics.

The object is to explain the effects of a d.c. electric field on a plane layer of dielectric liquid, and they observe that determining the onset of fluid motion and the consequent augmentation of charge transfer has led to a series of carefully controlled transient and steady-state electrochemical experiments and the application of theoretical instability analyses. By using ion-exchange membranes and electrodialytic varnishes on plane-parallel electrodes, it has been possible to investigate the consequences of almost space-charge-limited unipolar and bipolar injection into highly purified dielectric liquids and incorporate the modifying effects of residual conductivity. Deo & Richardson (1983) describe several experimental and linear instability results by P. Atten and his co-workers, working liquids being pyralenes, nitrobenzene, and propylene carbonate; they write p. 132, ... *because of the high degree of purity of the experimental liquids, the predictions of a diffusionless linear instability analysis that incorporates residual conductivity effects ... still do not appear to give a satisfactory explanation of the discrepancy between theory and experiment. All the theoretical estimates to date have arisen from instability analyses and have consistently exceeded the experimental values.* Presumably with this as a motivation, they chose to develop a generalized energy theory.

The precise problem studied by Deo & Richardson (1983) considers an incompressible dielectric liquid of constant density, ρ, electrical permittivity ϵ, and kinematic viscosity ν confined between two rigid planar perfectly conducting electrodes of infinite extent and distance d apart. They suppose an autonomous injection of unipolar charge is emitted from the electrode at $z = 0$, which is maintained at a potential Φ_0, and the collecting electrode at $z = d$ is maintained at zero potential. The governing electrodynamic field equations, neglecting magnetic effects, are

$$E_i = -\Phi_{,i}, \qquad D_{i,i} = Q, \qquad j_{i,i} = -\frac{\partial Q}{\partial t}, \qquad (11.23)$$

where Φ, E_i, D_i, j_i, and Q are electrical potential, electric field, electric displacement, current density, and space-charge density, respectively. The liquid is assumed to be a linear isotropic dielectric, charge is assumed to be transported by convection, migration and diffusion, and then the additional electrical constitutive equations adopted are

$$D_i = \epsilon E_i, \qquad j_i = Q u_i + K Q E_i - D_c Q_{,i}, \qquad (11.24)$$

where u_i is liquid velocity, D_c is the charge diffusion coefficient, and K is an ion mobility coefficient. The remaining equations are balance of mass and of linear momentum. Taking the electrical force to be Coulombic and the liquid to be an incompressible linear viscous one, these are

$$u_{i,i} = 0, \qquad (11.25)$$

$$u_{i,t} + u_j u_{i,j} = -\rho^{-1} p_{,i} + \nu \Delta u_i + \rho^{-1} Q E_i, \qquad (11.26)$$

where p is pressure.

Deo & Richardson (1983) next introduce the non-dimensional parameters

$$T = \frac{\epsilon \Phi_0}{\rho \nu K}, \qquad C = \frac{Q_0 d^2}{\epsilon \Phi_0}, \qquad M = \frac{1}{K} \left(\frac{\epsilon}{\rho}\right)^{1/2}, \qquad S = \frac{\epsilon D_c}{\rho \nu K^2}. \quad (11.27)$$

Here T represents the ratio of the Coulombic forces to viscous forces, C is a measure of the injected charge, M is the ratio of hydrodynamic mobility to ionic mobility, and S, which arises because charge-diffusion effects have been retained, is typically in the range 0.04–0.4. With this non-dimensionalization the governing equations admit a steady one-dimensional hydrostatic equilibrium solution with the equilibrium electric potential $\Phi_e(z)$ solving the equation

$$\frac{S}{T} \frac{d^4 \Phi_e}{dz^4} + \frac{1}{2} \frac{d^2}{dz^2} \left(\frac{d\Phi_e}{dz}\right)^2 = 0, \quad (11.28)$$

subject to the boundary conditions,

$$\Phi_e(0) = 1, \quad \Phi_e(1) = 0, \quad \frac{d^2}{dz^2} \Phi_e(0) = -C, \quad \frac{d^2}{dz^2} \Phi_e(1) = 0. \quad (11.29)$$

Even though Deo & Richardson (1983) are able to solve (11.28) numerically they observe that in most dielectric fluid experiments the parameter S/T is small and so $\Phi_e(z)$ exhibits a mainstream and boundary layer character. They also worked with the basic equilibrium field of mainstream character, which is

$$E_e(z) = \left(\frac{J^2}{C^2} + 2Jz\right)^{1/2}, \qquad Q_e(z) = \frac{J}{E_e(z)}, \quad (11.30)$$

J being a real constant uniquely determined from the boundary constraint

$$J^3 + (2C - 1)CJ^2 + \frac{4C^4 J}{3} - \frac{3C^4}{2} = 0. \quad (11.31)$$

Their energy stability analysis is carried out based on a perturbation to both solutions (11.28) and (11.30).

The resulting perturbations $u_i, e_i,$ and q in liquid velocity, electric field, and space-charge density satisfy the equations

$$u_{i,t} + \frac{T}{M^2} u_j u_{i,j} = -p_{,i} + \Delta u_i + T[qE_i + Qe_i + qe_i], \quad (11.32)$$

$$e_{i,t} = b_i - \frac{T}{M^2} \left[(Q + q)u_i + qE_i + (Q + q)e_i\right] + \frac{S}{M^2} q_{,i}, \quad (11.33)$$

$$q_{,t} + \frac{T}{M^2} u_j (Q + q)_{,j} = -\frac{T}{M^2} \left[qE_i + (Q + q)e_i\right]_{,i} + \frac{S}{M^2} \Delta q, \quad (11.34)$$

$$e_i = -\phi_{,i}, \qquad e_{i,i} = q, \qquad u_{i,i} = 0, \qquad (11.35)$$

where p, ϕ are perturbations in pressure and in potential, b_i is an arbitrary solenoidal vector field, and $\mathbf{E}$, Q refer to their values in (11.28). Deo & Richardson (1983) rely strongly on the fact that they consider a unipolar injection of charge so that

$$Q + q \geq 0, \qquad (11.36)$$

a condition, which means that nowhere in the liquid can the charge density change sign.

Deo & Richardson (1983) allow almost-periodic functions as perturbations and their analysis is based on a generalized energy of the form

$$E = \frac{1}{2} \left(\|\mathbf{u}\|^2 + M^2 [\gamma \|\mathbf{e}\|^2 + \lambda \|q\|^2] \right)$$

for $\gamma, \lambda (\geq 0)$ coupling parameters. Three choices are considered by Deo & Richardson (1983) and are called
   (i) a weighted energy functional for which $\lambda = 0$;
   (ii) a kinetic-charge functional for which $\gamma = 0$;
   (iii) a mixed functional in which $\gamma = 1$.
(It should be noted that this use of "weighted" energy does not correspond to that of chapter 6 where "weighted" implies use of a spatial weight.) As might be expected, the results for (iii) turn out to be sharpest.

We describe only the results for the "mixed" functional of case (iii). In general the analysis for case (i) is rigorous and leads to global monotonic decay although the results are somewhat conservative. The functional in case (ii) gives rise to a serious mathematical difficulty in bounding cubic terms and only conditional stability is achieved. The energy equation obtained for case (iii) is

$$\frac{dE}{dt} = T\big\{ < E_i q(u_i - e_i) > - < (Q+q)e_i e_i >$$
$$- \lambda < q(qE_i + Qe_i)_{,i} > -\frac{1}{2}\lambda < q^3 > -\lambda < qu_i Q_{,i} > \big\}$$
$$- \big[D(\mathbf{u}) + S\|q\|^2 + \lambda S D(q) \big]. \qquad (11.37)$$

The fundamental problem with this equation is the occurrence of the cubic terms $< qe_i e_i >$ and $< q^3 >$. Deo & Richardson (1983) overcome this in two ways; by first utilizing the constraint (11.36), and secondly by deriving a conditional stability result, i.e., for small enough initial data. (The constraint (11.36) is used in the manner

$$- < (Q+q)e_i e_i > -\frac{1}{2}\lambda < q^3 > \leq \frac{\lambda}{2} < Qq^2 > .)$$

Their criteria are based on three energy maximum quantities, $\overline{\overline{\tau}}_1$, which is defined by

$$\overline{\overline{\tau}}_1^{-1} = \min_{\lambda} \max_{\mathcal{H}} I/D, \qquad (11.38)$$

with

$$I = <E_i q(u_i - e_i)> -\lambda <q(qE_i + Qe_i)_{,i}>$$
$$- \lambda <qu_i Q_{,i}> + \frac{1}{2}\lambda <Qq^2>, \tag{11.39}$$

$$D = D(\mathbf{u}) + S\|q\|^2 + S\lambda D(q), \tag{11.40}$$

and $\overline{\overline{\tau}}_2$ and $\overline{\overline{\tau}}_\ell$. The quantity $\overline{\overline{\tau}}_\ell$ is the analogue of (11.38) for conditional stability. If $T < \overline{\overline{\tau}}_1$ or $T < \overline{\overline{\tau}}_\ell$ then nonlinear stability holds. The bound $\overline{\overline{\tau}}_2$ is an *estimate* of the *best value leaving in the cubic terms*. To derive $\overline{\overline{\tau}}_2$ they choose trial functions of the form

$$\phi = z^3(1-z)^3 \sin ax,$$

$$u_i = \big(2z(1-z)(1-2z)\cos ax, \, 0, \, az^2(1-z)^2 \sin ax\big),$$

although for this choice they observe the offending cubic terms integrate to zero. This heuristic procedure is not without interest; Deo & Richardson (1983) show that the ratio in the energy maximum problem for $\overline{\overline{\tau}}_1$ and $\overline{\overline{\tau}}_\ell$ is bounded above, but the existence question of an actual maximizing solution is not addressed. It may not be trivial in the class of almost periodic functions and is not at all clear for the quantity $\overline{\overline{\tau}}_2$.

The results of Deo & Richardson (1983) are certainly sharp, especially the global monotonic, $\overline{\overline{\tau}}_1$, values for case (iii). It is worth pointing out that the numerical procedures involve Chebyshev collocation techniques and are somewhat involved. This paper represents a substantial contribution to energy stability theory, especially since, as far as I can determine, it is the first to address the very interesting EHD problem.

12

Ferrohydrodynamic Convection

12.1 Introduction and the Basic Equations of Ferrohydrodynamics

Ferrohydrodynamics (FHD) is of great interest because the fluids of concern possess a giant magnetic response. This gives rise to several striking phenomena with important applications. Among these are the spontaneous formation of a labyrinthine pattern in thin layers, the self-levitation of an immersed magnet, and of particular interest here, the enhanced convective cooling in a ferrofluid that has a temperature-dependent magnetic moment. The very well written book by Rosensweig (1985) is a perfect introduction to this fascinating subject. He very briefly refers to thermo-convective instability in FHD, which is what we concentrate on here. Another, more general, but again very readable account of ferromagnetism may by found in Landau et al. (1984). We now present the relevant equations for FHD, in the forms appropriate to this chapter. Then a brief account is given of a striking convective-like instability, before embarking on the thermo-ferro convection problem.

Maxwell's equations are given below.

Faraday's law:

$$\nabla \times \mathbf{E} = -\frac{\partial \mathbf{B}}{\partial t}.$$

(12.1)

Ampère's law and Maxwell's correction:

$$\nabla \times \mathbf{H} = \mathbf{J} + \frac{\partial \mathbf{D}}{\partial t}.$$

(12.2)

Gauss' law (I):

$$\nabla . \mathbf{D} = Q;$$

(12.3)

Gauss' law (II):

$$\nabla . \mathbf{B} = 0.$$

(12.4)

Charge conservation:

$$\frac{\partial Q}{\partial t} + \nabla . \mathbf{J} = 0.$$

(12.5)

In these equations $\mathbf{E}$, $\mathbf{D}$, $\mathbf{H}$, $\mathbf{B}$ denote electric field, electric displacement, magnetic field, and magnetic induction, respectively. The quantities Q and $\mathbf{J}$ are free charge and current, respectively. In FHD, it is usual to assume the free charge Q and electric displacement $\mathbf{D}$ are absent. Hence, as Rosensweig (1985), p. 91, points out, the field equations (12.2), (12.4) are usually employed in the magnetostatic limit of Maxwell's equations,

$$\nabla \times \mathbf{H} = \mathbf{0}, \tag{12.6}$$

$$\nabla . \mathbf{B} = 0. \tag{12.7}$$

The magnetization $\mathbf{M}$ is introduced by

$$\mathbf{B} = \mu_0(\mathbf{H} + \mathbf{M}), \tag{12.8}$$

where μ_0 is the permeability of free space, $\mu_0 = 4\pi \times 10^{-7}$Henry m^{-1}.

In addition to the field equations (12.6), (12.7) a constitutive law is assumed relating $\mathbf{B}$ to $\mathbf{H}$. This has the form

$$\mathbf{B} = \mu(\mathbf{H}; T, \rho)\mathbf{H}, \tag{12.9}$$

where the T and ρ dependence are often suppressed.

The fluid is assumed incompressible and then with $\mathbf{v}$ denoting the velocity field, the continuity and momentum equations are

$$\nabla . \mathbf{v} = 0, \tag{12.10}$$

$$\frac{\partial \mathbf{v}}{\partial t} + (\mathbf{v}.\nabla)\mathbf{v} = -\frac{1}{\rho}\nabla p + \nu\Delta\mathbf{v} + \mathbf{f}, \tag{12.11}$$

where p is the pressure, ρ is the (constant) density, and $\mathbf{f}$ is the total force. In this work we allow $\mathbf{f}$ to be composed of two pieces, one due to gravity, while the other is due to the magnetic field. The gravity contribution may be written

$$-g\,\mathbf{f}_g, \tag{12.12}$$

where, for example, in a thermal convection problem,

$$\mathbf{f}_g = \left[1 - \alpha(T - T_R)\right]\mathbf{k}, \tag{12.13}$$

α being the thermal expansion coefficient, T temperature, T_R being a reference temperature, and $\mathbf{k} = (0, 0, 1)$.

The magnetic body force $\mathbf{f}_m$ has several forms and these are discussed in Rosensweig (1985), pp. 110–119, and in Landau et al. (1984), p. 127. For now we note that

$$\mathbf{f}_m = -\nabla\left[\mu_0 \int_0^H \left(\frac{\partial M \upsilon}{\partial \upsilon}\right)_{H,T} dH + \frac{1}{2}\mu_0 H^2\right] + \nabla(\mathbf{BH}) \tag{12.14}$$

or

$$\mathbf{f}_m = -\nabla\left[\mu_0 \int_0^H \left(\frac{\partial Mv}{\partial v}\right)_{H,T} dH\right] + \mu_0 M \nabla H, \tag{12.15}$$

where $v = 1/\rho$ is the specific volume and $M = |\mathbf{M}|$, $H = |\mathbf{H}|$. The above forms are useful in §12.2.

For convection problems, incorporating (12.13) and either (12.14) or (12.15) in (12.11), the momentum equation is conveniently written

$$\frac{d\mathbf{v}}{dt} = -\frac{1}{\rho}\nabla\tilde{p} + \nu\Delta\mathbf{v} - g\big(1 - \alpha[T - T_R]\big)\mathbf{k} + \nabla(\mathbf{BH}) \tag{12.16}$$

or

$$\frac{d\mathbf{v}}{dt} = -\frac{1}{\rho}\nabla\hat{p} + \nu\Delta\mathbf{v} - g\big(1 - \alpha[T - T_R]\big)\mathbf{k} + \mu_0 M\nabla H, \tag{12.17}$$

where the modified pressures $\tilde{p}, \hat{p}$ have the forms

$$\tilde{p} = p - \mu_0 \int_0^H \left(\frac{\partial Mv}{\partial v}\right)_{H,T} dH + \frac{1}{2}\mu_0 H^2, \tag{12.18}$$

$$\hat{p} = p - \mu_0 \int_0^H \left(\frac{\partial Mv}{\partial v}\right)_{H,T} dH. \tag{12.19}$$

Thus, the system of equations governing the ferrofluid motion are (12.6)–(12.10), together with either (12.16), (12.18) or (12.17), (12.19). A further equation for the temperature field must be added and this is discussed in §12.2.

This section is completed by reviewing a striking isothermal instability of Cowley & Rosensweig (1967), analysed nonlinearly by Gailitis (1977). Cowley & Rosensweig (1967) consider a horizontal layer of ferromagnetic fluid with a free surface. A magnetic field passes through the fluid at right angles, in the vertical direction. On the basis of a static linear instability theory they deduce that the free surface cannot remain flat and an instability must develop when the magnetic field strength exceeds a critical value, H_c. They also analysed the situation experimentally and verified their theoretical findings. Experimentally, the instability is truly striking. It consists of a very regular hexagon pattern of crests on the free surface, with spikes at the centre of the hexagon. Their working ferromagnetic fluid was prepared by grinding magnetite to particles of submicron size and adding to kerosene to which 5.8% by weight of oleic acid had been added. In a paper that demonstrates a very practical example of secondary bifurcation, Gailitis (1977) [it is worth pointing out that the paper was received on 19 March 1970!] further investigates the Cowley & Rosensweig (1967) problem. He used a potential energy argument (of the kind initiated by Kelvin (1887), as mentioned in chapter 1) and expanded the free surface and magnetic induction in a Fourier series. By examining various modes he discovered that a second critical field strength, H_2, exists, $H_2 > H_c$, such that the hexagonal shape exists for $H_c < H < H_2$, but after this square surface waveforms are the stable ones. The phenomenon of Gailitis (1977) exhibits hysteresis. To finish this section I think the papers of Cowley & Rosensweig (1967) and Gailitis (1977) are extremely fundamental in this field and are also very important mathematically.

12.2 Thermo-Convective Instability in FHD

Shliomis (1974) writes that the temperature dependence of the magnetization is important in thermo-convective instability in FHD. This point is, in fact, taken up in Finlayson (1970), Curtis (1971), Lalas & Carmi (1971), and Shliomis (1974). As we pointed out in the last section, the energy equation for the temperature field must be decided upon. For linear instability we neglect viscous dissipation in the energy equation and then the equation adopted by Finlayson (1970), and essentially by Curtis (1971) and Shliomis (1974) is

$$\left[\rho C_{VH} - \mu_0 \mathbf{H}.\left(\frac{\partial \mathbf{M}}{\partial T}\right)_{V,\mathbf{H}}\right] \frac{dT}{dt} + \mu_0 T \left(\frac{\partial \mathbf{M}}{\partial T}\right)_{V,\mathbf{H}} \cdot \frac{d\mathbf{H}}{dt} = k\Delta T, \quad (12.20)$$

where C_{VH} and k are heat capacity at constant volume and magnetic field and (constant) thermal conductivity, respectively. Thus, the complete system of equations is (12.20), (12.6)–(12.10) and *either* (12.16), (12.18) *or* (12.17), (12.19).

Finlayson (1970) takes

$$\mathbf{M} = \frac{\mathbf{H}}{H} M(H, T) \quad (12.21)$$

and

$$M = M_0 + \chi(H - H_0) - K(T - T_a), \quad (12.22)$$

where H_0 is a constant and T_a (constant) is an average temperature field. The coefficients χ and K are called the susceptibility and the pyromagnetic coefficient, respectively, and are defined by

$$\chi = \left(\frac{\partial M}{\partial H}\right)_{H_0, T_a}, \qquad K = -\left(\frac{\partial M}{\partial T}\right)_{H_0, T_a}. \quad (12.23)$$

The fluid is assumed to occupy the layer $z \in (-d/2, d/2)$, and the magnetic field, $\mathbf{H} = H_0^{ext} \mathbf{k}$, acts outside the layer.

Finlayson (1970) assumes

$$T = T_0 \quad \text{at} \quad z = \frac{1}{2}d, \qquad T = T_1 \quad \text{at} \quad z = -\frac{1}{2}d, \quad (12.24)$$

T_0, T_1 constants, and $T_a = \frac{1}{2}(T_0 + T_1)$. He investigates the instability of the solution

$$\mathbf{v} \equiv \mathbf{0}, \qquad T = T_a - \beta z,$$

$$\beta = \frac{T_1 - T_0}{d}, \qquad H_0 + M_0 = H_0^{ext},$$

$$\mathbf{H_0} = \mathbf{k}\left(H_0 - \frac{K\beta z}{1+\chi}\right), \qquad \mathbf{M_0} = \mathbf{k}\left(M_0 + \frac{K\beta z}{1+\chi}\right).$$

(12.25)

The analysis used is a linearized one based on stationary convection with the eigenvalue (critical Rayleigh number) being found numerically by a Galerkin method. In his work he introduces a constant C by

$$C = C_{VH} + \frac{\mu_0}{\rho}KH_0,$$

and his Rayleigh number is then

$$Ra = \frac{\alpha g \beta d^4}{\nu}\frac{\rho C}{k};$$

(12.26)

so the Rayleigh number includes the temperature dependence of the magnetization. His results are certainly interesting. In particular, he deduces that magnetic forces are only appreciably dominant in very thin layers; for example, in a 1mm layer of kerosene a temperature difference of 51°C is required to induce convection in a zero magnetic field, whereas in the same situation with the kerosene at saturation magnetization instability occurs for a temperature difference of only 19°C. For a depth of the order of 5mm, there is no difference.

The results of Curtis (1971) and Shliomis (1974) continue in this vein. Lalas & Carmi (1971) do attempt a different approach. They assume

$$M = M_0\big[1 - \gamma(T - T_R)\big],$$

(12.27)

and argue that for a constant magnetic field gradient

$$\nabla H > 100\,\text{Gauss/cm},$$

the terms involving the derivatives of $\mathbf{M}$ in (12.20) may be neglected thus reducing the energy equation to

$$\rho C_{VH}\frac{dT}{dt} = k\Delta T.$$

(12.28)

Their momentum equation becomes, using (12.27), (12.17), and (12.11),

$$\rho\frac{d\mathbf{v}}{dt} + \nabla\hat{p} = M_0\nabla H\big[1 - \gamma(T - T_R)\big] - \rho g\mathbf{k}\big[1 - \alpha(T - T_R)\big] + \rho\nu\Delta\mathbf{v}.$$

At this point they argue that the $M_0 \nabla H$ term can be regarded as constant and reduce the system to one equivalent to a standard Bénard one and can thus apply energy stability theory to derive *nonlinear* stability. To apply energy stability theory correctly it is surely necessary to include all perturbations in the $T \nabla H$ term. This problem would still appear to be open; the nonlinear result of Lalas & Carmi (1971) can only be regarded as heuristic. However, they define a modified Rayleigh number

$$Ra = \left| \alpha g + \frac{M_0 \gamma \nabla H}{\rho} \right| \frac{C_{VH} g \beta d^4}{k\nu}.$$

Provided $Ra < 1708$, they claim nonlinear stability.

To attempt to derive a rigorous nonlinear energy stability result we adopt an entirely different approach and return to the form for the magnetic force, (12.14) or (12.15).

12.3 Energy Stability for a Model Based on the Force Representation of Korteweg and Helmholtz

We now examine the possibility of developing a rigorous energy analysis for a material in which the permeability μ depends only on T, not on H. For this case the magnetic force assumes another form due to Korteweg and Helmholtz. A clear discussion of this is given by Rosensweig (1985), pp. 113, 114, and also Landau et al. (1984), p. 127. The relevant form is

$$\mathbf{f}_m = \nabla \left[\frac{H^2}{2} \rho \left(\frac{\partial \mu}{\partial \rho} \right)_T \right] - \frac{H^2}{2} \nabla \mu. \tag{12.29}$$

Rosensweig (1985) remarks that there has been some concern in the literature over this representation, but he expressly remarks (p. 118), ... *the Korteweg-Helmholtz expression stands out for its ability to explain experimental results in a straightforward way.*

The model we shall pursue is based on (12.6), (12.7),

$$\nabla \times \mathbf{H} = 0, \qquad \nabla . \mathbf{B} = 0, \tag{12.6, 12.7}$$

with the following form of (12.9)

$$\mathbf{B} = \mu(T)\mathbf{H}. \tag{12.30}$$

The equation we adopt for the temperature field is that of Carmi & Lalas (1971); namely, with $\kappa = k/\rho C_{VH}$,

$$\frac{\partial T}{\partial t} + v_i T_{,i} = \kappa \Delta T. \tag{12.31}$$

The continuity and momentum equations are as before with $\mathbf{f}_m$ given by (12.29), so

$$\nabla.\mathbf{v} = 0, \tag{12.32}$$

$$\frac{\partial v_i}{\partial t} + v_j v_{i,j} = -\omega_{,i} + \nu \Delta v_i - g\left[1 - \alpha(T - T_R)\right]k_i - \frac{1}{2}H^2 \mu_{,i}, \tag{12.33}$$

where now

$$\omega = p - \frac{1}{2}H^2 \rho \left(\frac{\partial \mu}{\partial \rho}\right)_T. \tag{12.34}$$

The model above is the ferromagnetic analogue of Roberts' (1969) first model discussed in §11.2.

To proceed we note that thanks to (12.6) we may introduce a potential, Φ, for $\mathbf{H}$ by

$$\mathbf{H} = -\nabla\Phi. \tag{12.35}$$

We also assume the explicit form for $\mu(T)$,

$$\mu = \mu_m\left[1 + \eta(T - T_m)\right], \tag{12.36}$$

where μ_m is the permeability at temperature T_m, and we select the temperature scale so that $T_m = 0$. We suppose the fluid is contained in the layer $z \in (0, d)$ and the boundaries are kept at fixed temperatures

$$T = T_d, \quad z = d; \qquad T = 0, \quad z = 0,$$

where, without loss of generality, the lower temperature is also selected as T_m. Then, the steady solution whose stability we examine is

$$\bar{T} = \beta z, \qquad \mathbf{v} = \mathbf{0},$$

with

$$\bar{\mu}\bar{H}_3 = \mu_m H_m \qquad \text{(constant)}.$$

Here $\bar{\mu} = \mu_m(1 + \eta\beta z)$ and so

$$\bar{H}_3 = -\frac{d\bar{\Phi}}{dz} = \frac{\mu_m H_m}{\bar{\mu}} = \frac{H_m}{1 + \eta\beta z}, \tag{12.37}$$

$\bar{\Phi}$ being the equilibrium magnetic potential. Furthermore, we deal only with the case where the fluid layer is *heated from above* so $T_d > 0$ and $\beta = T_d/d$.

Let the perturbed fields $\mathbf{H}$, Φ, T, and $\mathbf{v}$ be given by

$$\mathbf{H} = \bar{\mathbf{H}} + \mathbf{h}, \quad \Phi = \bar{\Phi} + \phi, \quad T = \bar{T} + \theta, \quad \mathbf{v} = \mathbf{0} + \mathbf{u}, \tag{12.38}$$

$\mathbf{h}, \phi, \theta, \mathbf{u}$ being (not necessarily small) perturbations. To proceed we must derive a suitable form of Boussinesq approximation and in this regard we follow Roberts' (1969) procedure for a dielectric fluid.

Equations (12.7) and (12.30) show

$$\nabla \cdot \left[(\bar{\mu} + \mu)(\bar{\mathbf{H}} + \mathbf{h}) \right] = 0,$$

where $\mu = \mu_m \eta \beta \theta$. Hence, since (12.7) also holds in equilibrium, this becomes

$$\eta(\theta \bar{\Phi}_{,z})_{,z} + \nabla \cdot \left[(1 + \eta T)\nabla \phi \right] = 0. \tag{12.39}$$

Roberts' (1969) argument is to assume $\eta T \ll 1$ and so discard any term involving ηT when a similar term occurs not containing that factor. Hence we drop the $\nabla \cdot (\eta T \nabla \phi)$ term in (12.39). Thus,

$$\begin{aligned} \Delta \phi &= -\eta(\theta \bar{\Phi}_{,z})_{,z} \\ &= \eta H_m (F(z)\theta)_{,z}, \end{aligned} \tag{12.40}$$

where $F(z) = 1/(1 + \eta \beta z)$. Again, using Roberts' hypothesis we drop the

$$(\eta \beta z \theta)_{,z} + \text{higher order terms}$$

from (12.40) to finally arrive at

$$\Delta \phi = \eta H_m \theta_{,z}. \tag{12.41}$$

It is of mathematical interest to note that we could attempt to keep a $\nabla \cdot (\eta \beta z \nabla \phi)$ term in (12.39), but to handle the nonlinear analysis that follows we would require an elliptic estimate for an equation of type (12.41) but with the Laplacian replaced by

$$\left(a_{ij}(z)\phi_{,j} \right)_{,i},$$

where a_{ij} is a diagonal matrix. I understand from Professor Galdi that there are counterexamples to this, and hence the physics employed in the Boussinesq approximation above is very important to the ensuing mathematics.

The perturbation temperature field, θ, satisfies

$$\theta_{,t} + \beta w + u_i \theta_{,i} = \kappa \Delta \theta, \tag{12.42}$$

where $w = u_3$. The velocity field $\mathbf{u}$ is again solenoidal and satisfies the equation, from (12.33),

$$\begin{aligned} u_{i,t} + u_j u_{i,j} = &- \omega_{,i} + \nu \Delta u_i + \alpha g k_i \theta \\ &- \frac{1}{2} \left[k_i \frac{\mu_m \eta \beta}{\rho} \{ 2\bar{\Phi}_{,z}\phi_{,z} + |\nabla \phi|^2 \} \right. \\ &\left. + \frac{\mu_m \eta \theta_{,i}}{\rho} \{ |\bar{\Phi}_{,z}|^2 + 2\bar{\Phi}_{,z}\phi_{,z} + |\nabla \phi|^2 \} \right], \end{aligned} \tag{12.43}$$

where ω now denotes the perturbation to the modified pressure. To proceed with the last term (in square brackets), we non-dimensionalize according to the scales,

$$\mathbf{u} = U\mathbf{u}^*, \qquad U = \nu/d, \qquad t = t^* d^2/\nu,$$

$$T^{\sharp} = U\sqrt{\frac{\beta\nu}{\alpha g\kappa}}, \qquad R^2 = \frac{\alpha\beta g d^4}{\kappa\nu}, \qquad \phi = A\phi^*,$$

$$\mathbf{x} = \mathbf{x}^* d, \qquad dH_m = A, \qquad \theta = \theta^* T^{\sharp},$$

where stars denote non-dimensional quantities. Then, omitting the stars, equation (12.43) non-dimensionalized is

$$u_{i,t} + u_j u_{i,j} = -\omega_{,i} + \Delta u_i + Rk_i\theta$$
$$- \frac{d^2}{2U\nu\rho}\left[k_i\beta\eta\mu_m\left\{-\frac{2H_m A}{d}F\phi_{,z} + \frac{A^2}{d^2}|\nabla\phi|^2\right\}\right]$$
$$- \frac{dT^{\sharp}\eta\mu_m}{2U\nu\rho}\left\{2\delta F^3 k_i H_m^2\theta\right.$$
$$\left. - \frac{2H_m A}{d}F\phi_{,z}\theta_{,i} + \frac{A^2}{d^2}\theta_{,i}|\nabla\phi|^2\right\}. \quad (12.44)$$

In arriving at this expression from (12.43) we have used (12.37), and, in particular, the term involving $\theta_{,i}|\bar{\Phi}_{,z}|^2$ has been written as a total derivative that is incorporated in the pressure, plus a piece in θ. The quantity F is defined by

$$F = \frac{1}{1 + \eta\beta dz}.$$

To complete the Boussinesq approximation we let $F \to 1$ in (12.44). It could be argued that we should also rewrite the term in $F\phi_{,z}\theta_{,i}$ as a term in θ, absorbing part of it in the pressure, but this term only plays a part in determining the size of the initial energy in a conditional analysis. It is not vital to the effect of stability, and so we leave it.

We also introduce the non-dimensional quantities

$$\delta = \beta d\eta, \qquad Pr = \nu/\kappa, \qquad L = \frac{\mu_m A^2}{\rho\nu^2},$$

and then equations (12.44), (12.42), and (12.41) may be reduced to

$$u_{i,t} + u_j u_{i,j} = -\omega_{,i} + \Delta u_i + R\theta k_i$$
$$+ \delta L k_i\left(\phi_{,z} - \delta\frac{Pr}{R}\theta\right) - \frac{1}{2}\delta L k_i|\nabla\phi|^2 \qquad (12.45)$$
$$+ L\delta\frac{Pr}{R}\phi_{,z}\theta_{,i} - \frac{1}{2}L\delta\frac{Pr}{R}\theta_{,i}|\nabla\phi|^2,$$

$$Pr(\theta_{,t} + u_i\theta_{,i}) = -Rw + \Delta\theta, \tag{12.46}$$

$$\Delta\phi = \delta\frac{Pr}{R}\theta_{,z}, \tag{12.47}$$

together with

$$u_{i,i} = 0. \tag{12.48}$$

We observe that the linearized system is mathematically equivalent to Roberts' (1969) first model. We do not discuss the linear instability analysis here, or the numerical results, together with their physical interpretation, since the purpose here is to demonstrate how a nonlinear energy stability theory may be developed for (12.45)–(12.48). However, numerical results for the linear and energy stability problems are presented for various boundary conditions in Straughan (1990).

The energy analysis we now present is mathematically analogous to one developed for Roberts' (1969) first model for a dielectric fluid by Galdi & Straughan (1990) where, as far as I am aware, elliptic estimates are applied in energy theory for the first time.

The boundary conditions we consider are that $\mathbf{u}, \theta, \phi$ have an x, y dependence, which forms a plane tiling periodic structure such as hexagons, and

$$\mathbf{u} = \mathbf{0}, \quad \theta = 0, \quad \phi = 0 \quad \text{on} \quad z = 0, d.$$

A wider boundary condition on ϕ, of the form

$$k\frac{\partial\phi}{\partial n} + a\phi = 0 \quad \text{on} \quad z = 0, d,$$

where a is a wavenumber, is used in Straughan (1990), where the linear instability boundary dependence on k is studied. This condition is analogous to (11.14), (11.15), used by Roberts (1969).

We emphasize that the novelty here is the use of elliptic estimates to control the nonlinearities in the energy analysis, and the particular choice of generalized energy necessary. We do not derive the elliptic estimates in detail, or indicate how they are valid for periodic (x, y) solutions. Further details are given, however, in Galdi & Straughan (1990).

Again, we note that we are treating the problem where the layer is heated from above. Denoting, as throughout the book, by $\|.\|, < . >$, the $L^2(V)$ norm and integral over V, V being a period-cell of the perturbation, we commence with the energy functional

$$E = \frac{1}{2}\|\mathbf{u}\|^2 + \frac{1}{2}Pr\|\theta\|^2.$$

(We could introduce a coupling parameter, but we concentrate on the aspect of how to control the nonlinearities and achieve a rigorous energy stability result.)

By differentiating E, using (12.45), (12.46), (12.48) and the boundary conditions we find

$$\frac{dE}{dt} = -D(\mathbf{u}) - D(\theta) + L\delta < \phi_{,z}w > -\delta^2 L \frac{Pr}{R} < \theta w > +\mathcal{N}. \quad (12.49)$$

The term $\mathcal{N}$ is given by

$$\mathcal{N} = I_1 + I_2 + I_3,$$

where

$$I_1 = -\frac{1}{2}\delta L < |\nabla\phi|^2 w >, \quad (12.50)$$

$$I_2 = L\delta\frac{Pr}{R} < \phi_{,z}\theta_{,i}u_i >, \quad (12.51)$$

$$I_3 = -\frac{1}{2}L\delta\frac{Pr}{R} < |\nabla\phi|^2\theta_{,i}u_i > . \quad (12.52)$$

We next derive an energy identity from (12.47) in the form

$$\|\nabla\phi\|^2 = -\delta\frac{Pr}{R} < \theta_{,z}\phi >, \quad (12.53)$$

and add this to (12.49), with a coupling parameter $\lambda \, (> 0)$, to obtain

$$\frac{dE}{dt} = -D + I + \mathcal{N}, \quad (12.54)$$

where

$$I = -\lambda\delta\frac{Pr}{R} < \theta_{,z}\phi > +L\delta < \phi_{,z}w > -\delta^2 L \frac{Pr}{R} < \theta w >,$$
$$D = D(\mathbf{u}) + D(\theta) + \lambda D(\phi). \quad (12.55)$$

With the energy E as given I do not see a way to control the nonlinearity $\mathcal{N}$ in (12.54). However, before introducing another piece, we present the elliptic estimates necessary. For the equation (12.47), we note

$$\Delta\phi = \delta\frac{Pr}{R} \, \text{div} \, (\theta\mathbf{k}).$$

With $\|.\|_r$ denoting the norm on $L^r(V)$ the appropriate estimate is

$$\|\nabla\phi\|_{2p} \le c_{2p}\|\theta\|_{2p}, \quad (12.56)$$

where c_{2p} is a constant depending on V. For example, from (12.53), we integrate by parts and then use the arithmetic-geometric mean inequality, for $\alpha \, (> 0)$ a constant to be chosen, to find

$$\|\nabla\phi\|^2 = \delta\frac{Pr}{R} < \theta\phi_{,z} > \le \delta\frac{Pr}{R} \left(\frac{1}{2\alpha}\|\phi_{,z}\|^2 + \frac{\alpha}{2}\|\theta\|^2 \right).$$

Choose $\alpha = \delta \frac{Pr}{R}$ and then

$$\|\nabla \phi\| \equiv \|\nabla \phi\|_2 \leq \delta \frac{Pr}{R} \|\theta\|.$$

To handle the nonlinearity $\mathcal{N}$ we deal with the terms $I_1 - I_3$ in turn.

$$
\begin{aligned}
< w|\nabla\phi|^2 > &\leq \|w\| < |\nabla\phi|^4 >^{1/2} \\
&\leq c_4^2 \|w\| \|\theta\|_4^2 \\
&\leq \tilde{c}_4 \|w\| D(\theta),
\end{aligned}
\tag{12.57}
$$

where the Cauchy-Schwarz inequality, (12.56) and then the Sobolev inequality $\|\theta\|_4 \leq cD^{1/4}(\theta)$ have been used. In (12.57) $\tilde{c}_4$ is a constant.

Next, by Hölder's inequality,

$$
\begin{aligned}
< \phi_{,z}\theta_{,i}u_i > &\leq \|\phi_{,z}\|_q < |\nabla\theta|^p|\mathbf{u}|^p >^{1/p}, & \frac{1}{p} + \frac{1}{q} = 1, \\
&\leq \|\phi_{,z}\|_q \|\nabla\theta\|_{ps} \|\mathbf{u}\|_{pr}, & \frac{1}{r} + \frac{1}{s} = 1.
\end{aligned}
$$

Pick $q = 6$, $p = 6/5$, $s = 5/3$, and $r = 5/2$. Then

$$
\begin{aligned}
I_2 &\leq L\delta\frac{Pr}{R} \|\nabla\phi\|_6 \|\nabla\theta\| \|\mathbf{u}\|_3 \\
&\leq \tilde{c} \|\theta\|_6 \|\nabla\theta\| \|\nabla\mathbf{u}\|,
\end{aligned}
\tag{12.58}
$$

where in the last line (12.56) and the Sobolev inequality $\|\mathbf{u}\|_3 \leq c\|\nabla\mathbf{u}\|$ have been employed.

Finally, using Hölder's inequality

$$
\begin{aligned}
< u_i\theta_{,i}|\nabla\phi|^2 > &\leq \|\nabla\phi\|_{2q}^2 < |\mathbf{u}|^p|\nabla\theta|^p >^{1/p}, & \frac{1}{p} + \frac{1}{q} = 1, \\
&\leq \|\nabla\phi\|_{2q}^2 \|\mathbf{u}\|_{pr} \|\nabla\theta\|_{ps}, & \frac{1}{r} + \frac{1}{s} = 1.
\end{aligned}
$$

Choose $p = 3/2$, $s = 4/3$, $r = 4$, and $q = 3$. Then using the Sobolev inequality $\|\mathbf{u}\|_6 \leq cD^{1/2}(\mathbf{u})$ we find

$$I_3 \leq \tilde{k}\|\theta\|_6^2 \|\nabla\mathbf{u}\| \|\nabla\theta\|. \tag{12.59}$$

Before using (12.57)–(12.59) in (12.54) we need to form another "energy" identity. Multiply (12.46) by θ^5 to obtain, with

$$E_1 = \frac{1}{6}Pr \int_V \theta^6 \, dV,$$

$$\frac{dE_1}{dt} = -R < \theta^5 w > -\frac{5}{9} \|\nabla\psi\|^2, \tag{12.60}$$

where $\psi = \theta^3$. We add (12.60) to (12.54) with the aid of a coupling parameter ξ, and utilize (12.57)–(12.59) to obtain

$$\frac{d}{dt}(E + \xi E_1) \leq I - D + \hat{c}_4 \|w\| D(\theta)$$
$$+ \tilde{c}\|\theta\|_6 \|\nabla\theta\| \|\nabla \mathbf{u}\| + \tilde{k}\|\theta\|_6^2 \|\nabla \mathbf{u}\| \|\nabla\theta\|$$
$$- \frac{5\xi}{9}\|\nabla\psi\|^2 + R\xi < w\theta^5 > . \tag{12.61}$$

The last term may be estimated with the aid of Hölder's inequality and the Sobolev inequality as

$$< w\theta^5 > \leq \|w\| < \psi^2|\psi||\theta| >^{1/2}$$
$$\leq \|w\| < \psi^6 >^{1/6} < |\psi|^{3/2}|\theta|^{3/2} >^{1/3}$$
$$\leq c\|w\| \|\nabla\psi\| \|\psi\|^{2/3}. \tag{12.62}$$

Hence, if we denote our generalized energy $\mathcal{E}$ by

$$\mathcal{E} = E + \xi E_1, \tag{12.63}$$

then

$$R\xi < w\theta^5 > \leq R\hat{c}\mathcal{E}^{1/3}\mathcal{D}, \tag{12.64}$$

for a computable constant $\hat{c}$, where $\mathcal{D}$ is defined by

$$\mathcal{D} = \zeta D + \frac{5\xi}{9}D(\psi), \tag{12.65}$$

for ζ a positive constant defined in (12.67).

The first two terms on the right of (12.61) are estimated as

$$I - D \leq -D\left(1 - \max_{\mathcal{H}} \frac{I}{D}\right),$$

where $\mathcal{H}$ is the space of admissible functions we work with. Define

$$R_E^{-1} = \max_{\mathcal{H}} \frac{I}{D}.$$

Then, we require

$$R_E^{-1} < 1. \tag{12.66}$$

This will be our nonlinear energy stability criterion. From (12.55) it obviously reduces to an eigenvalue problem involving L and R. Details of the resolution of this eigenvalue problem are not included here, but may be found in Straughan (1990).

Thus, suppose (12.66) holds and define

$$\zeta = 1 - R_E^{-1}(> 0). \tag{12.67}$$

We employ (12.63)–(12.65) in (12.61) to obtain

$$\begin{aligned}
\frac{d\mathcal{E}}{dt} &\leq -\mathcal{D} + c_1\mathcal{E}^{1/6}\mathcal{D} + c_2\mathcal{E}^{1/3}\mathcal{D} + c_3\mathcal{E}^{1/2}\mathcal{D} \\
&= -\mathcal{D}\big[1 - c_1\mathcal{E}^{1/6} - c_2\mathcal{E}^{1/3} - c_3\mathcal{E}^{1/2}\big],
\end{aligned} \tag{12.68}$$

where c_i are computable constants. If now

$$c_1\mathcal{E}^{1/6}(0) + c_2\mathcal{E}^{1/3}(0) + c_3\mathcal{E}^{1/2}(0) < 1, \tag{12.69}$$

and (12.66) are satisfied we may show from (12.68) that $\mathcal{E} \to 0$ as $t \to \infty$. We thus have a rigorous nonlinear energy stability result.

13

Convective Instabilities for Reacting Viscous Fluids Far from Equilibrium

13.1 Chemical Convective Instability

The phenomenon of double-diffusive convection in a fluid layer, where two scalar fields (such as heat and salinity concentration) affect the density distribution in a fluid, has become increasingly important in recent years. The behaviour in the double-diffusive case is much more diverse than for the Bénard problem. In particular, linear stability theory, cf. Baines & Gill (1969), predicts that the first occurrence of instability may be via oscillatory rather than stationary convection if the component with the smaller diffusivity is stably stratified. Finite amplitude convection in the double-diffusive context was investigated by Veronis (1965,1968a) whose results suggested steady finite amplitude motion could occur at critical values of a Rayleigh number much less than that predicted by linearized theory. Several later papers confirmed this, usually by weakly nonlinear theory, see e.g., Proctor (1981) and the references therein. Proctor's (1981) boundary layer analysis is an interesting one and provides some explanation for the energy results of Shir & Joseph (1968).

When temperature and one or more species are present and interactions between species are allowed, then the system becomes increasingly richer. Furthermore, reaction-diffusion equations for mixtures of viscous fluids play an important role in everyday life. We mention specifically acid rain effects, Pandis & Seinfeld (1989); the nuclear winter phenomenon, Giorgi (1989); warming of the stratosphere, Rood (1987), Kaye & Rood (1989); and enzyme recovery from reacting mixtures, Duong & Weiland (1981), Malikkides & Weiland (1982). However, the equations are usually written down in an ad hoc manner and often vary considerably. Therefore, a rational derivation of a relevant system of equations would seem appropriate. We therefore develop such a theory here and commence an analysis of linearized stability, following Morro & Straughan (1990). Before doing this, however, we examine a previous analysis of nonlinear energy stability in a dissociating gas.

13.2 Chemical Convective Instability and Quasi-Equilibrium Thermodynamics

Wollkind & Frisch (1971) developed a linear analysis for a chemical instability problem and deduced that for large enough Rayleigh number the Bénard problem involving a chemically dissociating fluid is unstable, in the situation that is stable for a non-reactive fluid, when the fluid layer is heated from above. The nonlinear analyses described for the Wollkind & Frisch system were given by Straughan et al. (1984).

The model with which Wollkind & Frisch commence considers a dissociating fluid contained in the infinite layer $0 < z < d$ and attention is restricted to essentially isochoric motions by adopting a Boussinesq-type approximation. The mass flux through the boundaries $z = 0, d$ is zero, and the prescribed temperatures there are T_0, T_1, with $T_1 > T_0$. The relevant equations admit a steady solution in which the velocity is zero, the fraction, α, of free atoms is constant, and the temperature, T, is linear in z across the layer.

In non-dimensional form the equations for the perturbation to the constant concentration solution are

$$\dot{u}_i = -p_{,i} + \delta_{i3}(R\theta + \phi) + \Delta u_i,$$
$$Sc\,\dot{\phi} = \Delta\phi - X_1\phi, \tag{13.1}$$
$$Pr\,\dot{\theta} = -Rw + \Delta\theta + \left(S + \frac{PrA}{ScR}\right)\Delta\phi - \frac{X_1 PrA}{ScR}\,\phi,$$

where $\mathbf{u}, \theta, \phi, p$ are the perturbation fields of velocity (solenoidal), temperature, fraction of free atoms, and pressure, R^2 is the Rayleigh number, Δ is the three-dimensional Laplacian, a superposed dot denotes the material derivative, $w = u_3$, and where for completeness we include below the non-dimensionalization appropriate to the notation of Wollkind & Frisch (1971),

$$Sc = \nu/D_{12}^0, \qquad X_1 = d^2/\tau D_{12}^0, \qquad \kappa = \lambda^0/\rho_0 c_1,$$
$$Pr = \nu/\kappa, \qquad S = D_{12}^0 D_0/2mc_1\kappa, \qquad A = c_2\alpha/c_1 b.$$

The functions $\mathbf{u}, \theta, \phi, p$ are assumed periodic in x, y and satisfy the boundary conditions

$$\mathbf{u} = \mathbf{0}, \qquad \frac{\partial\phi}{\partial z} = 0, \qquad \theta = 0, \tag{13.2}$$

on $z = 0, d$.

To investigate nonlinear stability we now define, for $\lambda\,(> 0)$ to be chosen, an energy $E(t)$ by

$$E(t) = \frac{1}{2}\left(\|\mathbf{u}\|^2 + Pr\|\theta\|^2 + \lambda Sc\|\phi\|^2\right), \tag{13.3}$$

where for V a period cell of the perturbed solution, $\| \cdot \|$, $< \cdot >$, and $D(\cdot)$, denote again the L^2-norm, integral, and Dirichlet integral on V.

The independence of equation $(13.1)_2$ allows us to use λ in a different way from the normal theory of the best λ. From (13.1) we derive

$$
\frac{dE}{dt} = < \phi w > -D(\mathbf{u}) - \lambda D(\phi) - D(\theta) - \lambda X_1 \|\phi\|^2
$$
$$
- \left(S + \frac{PrA}{ScR} \right) < \nabla\phi . \nabla\theta > - \frac{X_1 PrA}{ScR} < \phi\theta > . \tag{13.4}
$$

From use of the arithmetic-geometric mean and Poincaré inequalities we next establish the following estimates:

$$
< \phi w > \le \frac{1}{2} \lambda_1 \|w\|^2 + \frac{1}{2} \lambda_1^{-1} \|\phi\|^2,
$$

$$
- \left(S + \frac{PrA}{ScR} \right) < \nabla\phi . \nabla\theta > \le \frac{1}{3} D(\theta) + \frac{3}{4} \left(S + \frac{PrA}{ScR} \right)^2 D(\phi),
$$

$$
\frac{X_1 PrA}{ScR} < \phi\theta > \le \frac{1}{3} D(\theta) + \frac{3}{4\lambda_1} \left(\frac{X_1 PrA}{ScR} \right)^2 \|\phi\|^2,
$$

where λ_1 is the constant in Poincaré's inequality. These estimates are used in (13.4), and we then select λ so large that

$$
\lambda \ge \frac{3}{4} \left(S + \frac{PrA}{ScR} \right)^2
$$

and

$$
\lambda X_1 \ge \lambda_1^{-1} + \frac{3}{2\lambda_1} \left(\frac{X_1 PrA}{ScR} \right)^2,
$$

and from the resulting inequality and further use of Poincaré's inequality we obtain

$$
\frac{dE}{dt} \le -ME,
$$

where

$$
M = \min \left\{ \lambda_1, 2\lambda_1/3Pr, X_1/Sc \right\}.
$$

Clearly then, $E \to 0$ as $t \to \infty$, and so there is *no* instability when the fluid layer is heated from above. (Such a conclusion was suggested from a different viewpoint by Wollkind & Bdzil (1971).)

We now include an analysis of another system suggested by Wollkind & Frisch (1971).

A perturbation to the constant concentration equilibrium solution for the *modified* system of Wollkind & Frisch (1971) (using *their* chemical quasi-equilibrium approximation) satisfies the equations

$$
\begin{aligned}
\dot{u}_i &= -p_{,i} + g\alpha k_i \theta + gbk_i \phi + \nu \Delta u_i, \\
\dot{\theta} &= \beta w + \kappa \Delta \theta, \\
\dot{\theta} &= \beta w - M\phi + D_{12}^0 \Delta \theta,
\end{aligned}
\tag{13.5}
$$

for divergence free **u**. The constant coefficents are in the notation of Wollkind & Frisch (1971) and we do not include them explicitly as we give a non-dimensional version, which corresponds to our notation below. A key factor is that from $(13.5)_{2,3}$

$$
\phi = \Delta\theta \left[\frac{D_{12}^0 - \kappa}{M} \right],
$$

and hence ϕ may be eliminated to yield the following (non-dimensional) equations,

$$
\begin{aligned}
\dot{u}_i &= -p_{,i} + \delta_{i3} R(\theta + \epsilon B \Delta\theta) + \Delta u_i, \\
Pr\dot{\theta} &= -Rw + \Delta\theta,
\end{aligned}
\tag{13.6}
$$

in which ϵB is a reaction term of small magnitude, introduced by Wollkind & Frisch (1971).

To investigate the stability of the zero solution to (13.6) we choose

$$
E(t) = \frac{1}{2}\|\mathbf{u}\|^2 + \frac{1}{2}Pr\|\theta\|^2.
\tag{13.7}
$$

The energy equation appropriate to (13.7) is determined to be

$$
\frac{dE}{dt} = RI - \mathcal{D},
\tag{13.8}
$$

where

$$
\mathcal{D} = D(\mathbf{u}) + D(\theta), \qquad I = -\epsilon B < \nabla\theta.\nabla w > .
$$

Define now

$$
\frac{1}{R_E} = \max \frac{I}{\mathcal{D}} \quad (= \lambda),
\tag{13.9}
$$

where the maximum is over the space of admissible solutions and from (13.8)

$$
\frac{dE}{dt} \leq -\mathcal{D}R\left(\frac{1}{R} - \frac{1}{R_E}\right).
\tag{13.10}
$$

If now $R < R_E$, then (13.10) and Poincaré's inequality show that $E \to 0$ at least exponentially as $t \to \infty$.

The problem is then to find R_E, or equivalently λ, as in (13.9). To this end we derive the Euler-Lagrange equations for this maximum as

$$\epsilon B \delta_{i3} \Delta \theta + 2\lambda \Delta u_i = 2p_{,i},$$
$$\epsilon B \Delta w + 2\lambda \Delta \theta = 0. \tag{13.11}$$

These equations are linear and so we may use a normal mode technique to obtain

$$4\lambda^2 (D^2 - a^2)^2 W = -a^2 (\epsilon B)^2 (D^2 - a^2)W, \tag{13.12}$$

where $D = d/dz$, a^2 is the wave number, and $W(z)$ is the z-part of w. For the two free boundaries situation covered in Wollkind & Frisch (1971) the boundary conditions allow W to be composed of $\sin m\pi z$, $m = 1, 2, ...$, and (13.12) yields

$$\lambda^2 = \frac{(\epsilon B)^2 a^2}{4(m^2 \pi^2 + a^2)}. \tag{13.13}$$

Obviously, as a function of m, λ is maximum for $m = 1$. We then see that the maximum of λ is achieved asymptotically as $a^2 \to \infty$. We may, therefore, conclude that $R_E = 2/\epsilon B$.

If we denote by R_L^2 the critical Rayleigh number of linear theory, the asymptotic expression given in Wollkind & Frisch (1971), eq. (36), is

$$R_L^2 = \frac{4}{(\epsilon B)^2} + \frac{2\pi^2}{\epsilon B} + O(1), \tag{13.14}$$

which compares with the energy limit

$$R_E^2 = \frac{4}{(\epsilon B)^2}. \tag{13.15}$$

Estimates (13.14) and (13.15), which agree to leading order, determine quantitatively a band of Rayleigh numbers where subcritical bifurcation may occur.

13.3 Basic Equations for a Chemically Reacting Mixture

General mixture theories have been considered by several writers; a lucid description is contained in the book by Müller (1985).

In this and the next two sections we describe a theory of a mixture of chemically reacting viscous fluids due to Morro & Straughan (1990) that uses one velocity field. They argue that since even the study of convection problems in this case requires substantial numerical computation this justifies the approach since any attempt to study convection in reacting fluids with individual velocities will be difficult, cf. Morro & Romeo (1988). Also, a study of convection in a multicomponent reacting mixture is very relevant at this time, partly because of the applications outlined earlier but also since multicomponent studies with no reaction effects are simultaneously attracting attention in the literature, see e.g. Pearlstein et al. (1989) and the references therein.

Consider a mixture of $N + 1$ fluids, with each constituent labelled by a Greek index. The notation we employ is standard; a repeated Greek index signifies summation from 1 to N whereas a repeated Roman index denotes summation over 1 to 3. We denote by ρ_α the mass density of the αth constituent, and by $\rho\, (= \sum_{\alpha=1}^{N+1} \rho_\alpha)$ the total density; $c_\alpha = \rho_\alpha/\rho$ is the concentration of the αth species, $\mathbf{h}_\alpha$ is the relative αth mass flux, and m_α is the mass supply.

The equation of balance of mass for each constituent is

$$\rho\dot{c}_\alpha = -\nabla \cdot \mathbf{h}_\alpha + m_\alpha, \qquad \alpha = 1, ..., N, \tag{13.16}$$

where a superposed dot denotes the material time derivative; in (13.16) the $N + 1$th components are determined from the relations

$$\sum_{\alpha=1}^{N+1} c_\alpha = 1, \qquad \sum_{\alpha=1}^{N+1} \mathbf{h}_\alpha = 0, \qquad \sum_{\alpha=1}^{N+1} m_\alpha = 0.$$

The equations of balance of mass, linear momentum, and energy for the total mixture are, respectively,

$$\dot{\rho} + \rho\nabla \cdot \mathbf{v} = 0, \tag{13.17}$$

$$\rho\dot{\mathbf{v}} = \nabla \cdot \mathbf{T} + \rho\mathbf{b}, \tag{13.18}$$

$$\rho\dot{e} = -\nabla \cdot \mathbf{q} + \mathbf{T} \cdot \mathbf{D} + \rho r, \tag{13.19}$$

where $\mathbf{v}$ is the velocity, $\mathbf{T}$ is the stress tensor, $\mathbf{b}$ is the body force, e is the internal energy, $\mathbf{q}$ is the heat flux, $\mathbf{D}$ is the symmetric part of the velocity gradient $\mathbf{L}$, and r is the heat supply.

Define η to be the entropy and θ the temperature. Morro & Straughan (1990) employ the entropy inequality in the form

$$\rho\dot{\eta} \geq -\nabla \cdot \left(\frac{\mathbf{q}}{\theta} + \mathbf{k}\right) + \frac{\rho r}{\theta}, \tag{13.20}$$

where $\mathbf{k}$ is Müller's entropy extra flux. For application, this inequality is rewritten in terms of the free energy ψ, as

$$-\rho(\dot{\psi} + \eta\dot{\theta}) + \mathbf{T} \cdot \mathbf{D} - \frac{\mathbf{q} \cdot \mathbf{g}}{\theta} - \theta\nabla \cdot \mathbf{k} \geq 0, \tag{13.21}$$

$\mathbf{g}$ denoting the temperature gradient.

Consequences from the relations (13.16)–(13.21) are now derived for an incompressible mixture, that is a mixture of $N+1$ fluids with ρ constant and so equation (13.17) becomes

$$\nabla \cdot \mathbf{v} = 0. \tag{13.22}$$

Morro & Straughan (1990) adopt the following constitutive theory:

$$\psi, \mathbf{T}, \eta, \mathbf{q}, \mathbf{k}, \mathbf{h}_\alpha, \text{ and } m_\alpha$$

are functions of

$$\theta, C, \mathbf{g}, \mathbf{S}, \mathbf{D},$$

where $C = (c_1, ..., c_N)$, $\mathbf{S} = (\mathbf{s}_1, ..., \mathbf{s}_N)$, with $\mathbf{s}_\alpha = \nabla c_\alpha$. Then setting $\mu_\alpha = \partial \psi / \partial c_\alpha$ and $\mathbf{J} = \theta \mathbf{k} + \mu_\alpha \mathbf{h}_\alpha$ they write the entropy inequality as

$$
\begin{aligned}
- \rho(\psi_\theta + \eta)\dot\theta &- \rho\psi_{\mathbf{g}} \cdot \dot{\mathbf{g}} - \rho\psi_{\mathbf{s}_\alpha} \cdot \dot{\mathbf{s}}_\alpha - \rho\psi_{\mathbf{D}} \cdot \dot{\mathbf{D}} \\
&+ \mathbf{T} \cdot \mathbf{D} + \left(\mathbf{J}_\theta - \frac{\mathbf{q}}{\theta} - \mathbf{k} - \mathbf{h}_\alpha \frac{\partial \mu_\alpha}{\partial \theta}\right) \cdot \mathbf{g} \\
&+ \left(\mathbf{J}_{c_\alpha} - \mathbf{h}_\beta \frac{\partial \mu_\beta}{\partial c_\alpha}\right) \cdot \mathbf{s}_\alpha + \mathbf{J}_{\mathbf{g}} \cdot (\nabla\nabla\theta) \\
&+ \mathbf{J}_{\mathbf{s}_\alpha} \cdot (\nabla\nabla c_\alpha) + \mathbf{J}_{\mathbf{D}} \cdot (\nabla \mathbf{D}) - \mu_\alpha m_\alpha \geq 0. \tag{13.23}
\end{aligned}
$$

They make use of standard thermodynamic arguments and the fact that $\dot\theta, \dot{\mathbf{g}}, \dot{\mathbf{s}}_\alpha, \dot{\mathbf{D}}$ appear linearly in (13.23) to deduce the facts below.

The free energy ψ is independent of $\mathbf{g}, \mathbf{S}, \mathbf{D}$, and so,

$$\psi = \psi(\theta, C).$$

The entropy η is related to the free energy ψ by

$$\eta = -\psi_\theta. \tag{13.24}$$

The quantity $\mathbf{J}$, which is the energy flux due to diffusion, satisfies the restrictions

$$\text{sym}\, \mathbf{J}_{\mathbf{g}} = 0, \qquad \text{sym}\, \mathbf{J}_{\mathbf{s}_\alpha} = 0, \quad \alpha = 1, ..., N, \qquad \mathbf{J}_{\mathbf{D}} = 0, \tag{13.25}$$

where sym denotes the symmetric part.

The entropy inequality remaining from (13.23) is

$$
\begin{aligned}
\mathbf{T} \cdot \mathbf{D} &+ \left(\mathbf{J}_\theta - \frac{\mathbf{q}}{\theta} - \mathbf{k} - \mathbf{h}_\alpha \frac{\partial \mu_\alpha}{\partial \theta}\right) \cdot \mathbf{g} \\
&+ \left(\mathbf{J}_{c_\alpha} - \mathbf{h}_\beta \frac{\partial \mu_\beta}{\partial c_\alpha}\right) \cdot \mathbf{s}_\alpha - m_\alpha \mu_\alpha \geq 0, \tag{13.26}
\end{aligned}
$$

and this must be true for every motion of the incompressible mixture.

Morro & Straughan (1990) appeal to a result of Gurtin (1971), to conclude that for $\mathbf{J}$ isotropic, (13.25) implies it vanishes. Thus when $\mathbf{J} = 0$ the entropy inequality finally becomes

$$\mathbf{T} \cdot \mathbf{D} - \left[\frac{\mathbf{q}}{\theta} + \left(\frac{\partial \mu_\alpha}{\partial \theta} - \frac{\mu_\alpha}{\theta} \right) \mathbf{h}_\alpha \right] \cdot \mathbf{g} - \mathbf{h}_\beta \frac{\partial \mu_\beta}{\partial c_\alpha} \cdot \mathbf{s}_\alpha - \mu_\alpha m_\alpha \geq 0, \quad (13.27)$$

and the entropy flux $\mathbf{k}$ has form

$$\mathbf{k} = -\frac{\mu_\alpha \mathbf{h}_\alpha}{\theta}.$$

13.4 A Model for Reactions Far from Equilibrium

To apply the general theory above it is necessary to be more specific about the form of constitutive equations. Before doing this, however, it is instructive to the understanding of the problem to review some basic properties of chemistry connecting the chemical affinity of a reaction and the chemical reaction rates.

Chemical Affinities and Mass Supplies. Let the $N+1$ molecular constituents involved in the chemical reactions be composed of A atomic substances, where $A \leq N$. For $a = 1, ..., A$ and $\beta = 1, ..., N + 1$, let $T_{a\beta}$ represent the number of atoms of the ath species in the molecule of the βth species. For the case under consideration the atomic substances are assumed indestructabile, and so

$$\sum_{\beta=1}^{N+1} T_{a\beta} \frac{m_\beta}{M_\beta} = 0, \qquad a = 1, ..., A, \qquad (13.28)$$

M_β being the molecular weight. As a consequence of (13.28), and since $\varrho = \text{rank}\,(T_{a\beta}) \leq A$, the reaction rates j_r, $r = 1, ..., N + 1 - \varrho$, determine the mass supplies m_β through the equations

$$m_\beta = \rho M_\beta \sum_r P_{\beta r} j_r, \qquad (13.29)$$

where $P_{\beta r}$ are the stoichiometric coefficients. These coefficients are not totally independent since the total mass supply is zero, i.e. $\sum_{\beta=1}^{N+1} m_\beta = 0$. This follows because if $\nu_{\beta r} = \rho M_\beta P_{\beta r}$, we may sum (13.29) over β and then because of the arbitrariness of j_r one sees that

$$\sum_{\beta=1}^{N+1} \nu_{\beta r} = 0, \qquad r = 1, ..., N + 1 - \varrho. \qquad (13.30)$$

It may be shown that the chemical affinity of reaction r, A_r, is given in terms of the reduced chemical potentials μ_α by

$$A_r = \nu_{\alpha r}\mu_\alpha, \tag{13.31}$$

where summation is over $\alpha = 1, ..., N$. This then leads to the useful relation connecting mass supplies and reaction rates,

$$m_\alpha\mu_\alpha = \sum_r j_r A_r. \tag{13.32}$$

Thermodynamic Deductions. Morro & Straughan (1990) assume that the nonlinearities of the model occur only through the mass supplies. They follow the approach of Loper & Roberts (1978,1980) in compositional convection and introduce the explicit (linear) constitutive assumptions:

$$\begin{aligned}
\psi &= \psi(\theta, C), \\
\mathbf{q} &= -\kappa\mathbf{g} - \omega_\alpha\mathbf{s}_\alpha, \\
\mathbf{h}_\alpha &= \chi_\alpha\mathbf{g} - \lambda_{\alpha\beta}\mathbf{s}_\beta, \\
\mathbf{T} &= -p\mathbf{I} + 2\mu\mathbf{D},
\end{aligned} \tag{13.33}$$

where $\kappa, \omega_\alpha, \chi_\alpha, \lambda_{\alpha\beta}, p, \mu$ are functions of θ, C; the mass supplies $m_\alpha(= m_\alpha(\theta, C))$ are still general functions of their arguments but they may be written in terms of the reaction rates j_r according to (13.32). Upon substitution into the entropy inequality (13.27) they deduce that

$$\mu > 0,$$

and the residual entropy inequality yields

$$\frac{\mathbf{q} \cdot \mathbf{g}}{\theta} + \mu_\alpha m_\alpha + \zeta_\alpha\mathbf{h}_\alpha \cdot \mathbf{g} + a_{\alpha\beta}\mathbf{s}_\alpha \cdot \mathbf{h}_\beta \leq 0, \tag{13.34}$$

where $\zeta_\alpha = \partial\mu_\alpha/\partial\theta - \mu_\alpha/\theta$, $a_{\alpha\beta} = \partial\mu_\alpha/\partial c_\beta$, and the matrix $(a_{\alpha\beta})$ is symmetric. Upon employing equations $(13.33)_{2,3}$ in (13.34) one obtains

$$\left(-\frac{\kappa}{\theta} + \zeta_\alpha\chi_\alpha\right)\mathbf{g} \cdot \mathbf{g} - a_{\gamma\beta}\lambda_{\gamma\alpha}\mathbf{s}_\alpha \cdot \mathbf{s}_\beta + \left(\chi_\alpha a_{\alpha\beta} - \zeta_\alpha\lambda_{\alpha\beta} - \frac{\omega_\beta}{\theta}\right)\mathbf{s}_\beta \cdot \mathbf{g} + \mu_\alpha m_\alpha \leq 0.$$

In view of the fact that $\mu_\alpha m_\alpha$ is independent of $\mathbf{g}$, $\mathbf{s}_\beta$, Morro & Straughan (1990) then conclude that this inequality can hold if and only if

$$\mu_\alpha m_\alpha \leq 0,$$

or, alternatively, due to (13.32)

$$\sum_r j_r A_r \leq 0$$

and, in addition, the matrix

$$
\begin{pmatrix}
\kappa/\theta - \chi_\gamma \zeta_\gamma & \frac{1}{2}(\zeta_\gamma \lambda_{\gamma\alpha} - \chi_\gamma a_{\gamma\alpha} + \omega_\alpha/\theta) \\
\frac{1}{2}(\zeta_\gamma \lambda_{\gamma\beta} - \chi_\gamma a_{\gamma\beta} + \omega_\beta/\theta) & a_{\gamma\alpha} \lambda_{\gamma\beta}
\end{pmatrix}
$$

is positive semidefinite.

Reduction of the energy equation. The energy equation (13.19) is still too general for practical purposes and some reduction is necessary. This is done by recalling the definition of μ_α and using (13.16) and (13.24) to rewrite equation (13.19) as

$$
\rho \theta \eta_\theta \dot{\theta} - \rho \theta \frac{\partial \mu_\alpha}{\partial \theta} \dot{c}_\alpha = -\nabla \cdot (\mathbf{q} - \mu_\alpha \mathbf{h}_\alpha) + \mathbf{T} \cdot \mathbf{D} - \mathbf{h}_\alpha \cdot \nabla \mu_\alpha - m_\alpha \mu_\alpha. \quad (13.35)
$$

Without loss of physical content, we now set r equal to zero. The terms $\mathbf{T}$ and $\mathbf{h}_\alpha$ are expressed using $(13.33)_{3,4}$ to find

$$
\mathbf{T} \cdot \mathbf{D} - \mathbf{h}_\alpha \cdot \nabla \mu_\alpha = 2\mu \mathbf{D} \cdot \mathbf{D} + \left(\frac{\partial \mu_\alpha}{\partial \theta} \mathbf{g} + \frac{\partial \mu_\alpha}{\partial c_\omega} \mathbf{s}_\omega \right) \cdot (\lambda_{\alpha\beta} \mathbf{s}_\beta - \chi_\alpha \mathbf{g}). \quad (13.36)
$$

By following an argument similar to that of chapter 3 we neglect the terms of (13.36) in (13.35), which all involve products of gradients. This leads to a reduced energy equation from (13.35) of form

$$
\rho \theta \eta_\theta \dot{\theta} - \rho \theta \frac{\partial \mu_\alpha}{\partial \theta} \dot{c}_\alpha = -\nabla \cdot (\mathbf{q} - \mu_\alpha \mathbf{h}_\alpha) - m_\alpha \mu_\alpha. \quad (13.37)
$$

Selection of m_α and ψ. Up to this point linear constitutive equations (in $\mathbf{g}$ and $\mathbf{s}_\alpha$) have been prescribed but m_α has not been specified. When we wish to analyze a specific problem it is then necessary to prescribe equations for m_α and ψ. This approach is not unlike that of Loper & Roberts (1980), p. 91, who derive an exact model to describe compositional convection.

Morro & Straughan (1990) choose the free energy ψ to be

$$
\psi = \tilde{c}\theta \left\{ 1 - \log\left(\frac{\theta}{\tilde{\theta}} \right) \right\} + \epsilon_\alpha c_\alpha \theta + \frac{1}{2} \phi_{\alpha\beta} c_\alpha c_\beta, \quad (13.38)
$$

where $\tilde{c}, \epsilon_\alpha, \phi_{\alpha\beta}, \tilde{\theta}$ are constants. They further assume a linear relationship between the mass supplies and the temperature and concentrations so that for constants $\tau_{\alpha\beta}, \delta_\alpha$,

$$
m_\alpha = -\tau_{\alpha\beta} c_\beta - \delta_\alpha \theta. \quad (13.39)
$$

Since the entropy inequality (13.34) requires $\mu_\alpha m_\alpha \le 0$, this implies $\tau_{\alpha\beta}, \delta_\alpha$, $\epsilon_\alpha, \phi_{\alpha\beta}$ must comply with the restriction

$$
(\tau_{\alpha\beta} c_\beta + \delta_\alpha \theta)(\epsilon_\alpha \theta + \phi_{\alpha\beta} c_\beta) \ge 0, \quad (13.40)
$$

for all admissible temperatures and concentrations θ, c_α.

The energy balance equation (13.37) and the balance of mass laws (13.16) are then succinctly written with the aid of (13.33) and (13.38) as

$$\rho c\dot\theta - \rho\epsilon_\alpha\theta\dot c_\alpha = \nabla \cdot (k\mathbf{g}) + \nabla \cdot (l_\alpha\mathbf{s}_\alpha) - m_\alpha\mu_\alpha, \qquad (13.41)$$

$$\rho\dot c_\alpha = \nabla \cdot (\lambda_{\alpha\beta}\mathbf{s}_\beta) - \nabla \cdot (\chi_\alpha\mathbf{g}) + m_\alpha, \qquad (13.42)$$

where $k = \kappa + \mu_\alpha\chi_\alpha$, and $l_\alpha = \lambda_\alpha - \mu_\gamma\lambda_{\gamma\alpha}$.

To complete the model it is necessary to prescribe a form for the buoyancy term in the momentum equation. Morro & Straughan (1990) employ a Boussinesq approximation and the momentum equation then becomes

$$\dot v_i = -\frac{1}{\rho}p_{,i} + \nu\Delta v_i - g\delta_{i3}\big[1 - \alpha(\theta - \bar\theta) + \beta_\omega(c_\omega - \bar c_\omega)\big], \qquad (13.43)$$

where g is the gravity acceleration, $\rho, \nu, \alpha, \beta_\omega$ are constants, and $\bar\theta, \bar c_\omega$ are conveniently chosen, constant reference values.

The model of convection in a chemically reacting mixture, not necessarily near equilibrium, is then composed of the system of equations (13.41)–(13.43) together with the incompressibility condition (13.22). For any model to be practically useful it must allow the determination of the solution or certainly it must allow deduction of useful quantitative information. The model described above is still highly nonlinear and numerical computation is evidently necessary.

To investigate convective instability, Morro & Straughan (1990) resort to a yet simpler version of (13.41)–(13.43). They suppose $\lambda_{\alpha\beta}$ and k are constant and take $l_\alpha, \chi_\alpha = 0$, and thereby neglect, respectively, the Dufour and Soret effects. (Within the context of a non-reacting mixture such effects have been intensely investigated, see e.g., Ybarra & Velarde (1979) and the references therein.) The inclusion of such effects may, however, be desirable in a detailed study of a specific reacting gaseous atmosphere.

13.5 Convection in a Layer

Suppose now the reacting mixture is confined to the layer $z \in (0, d)$, with the boundaries $z = 0, d$ retained at fixed temperatures, namely,

$$\theta = \theta_1, \quad z = 0; \qquad \theta = \theta_2, \quad z = d, \qquad (13.44)$$

with $\theta_2 < \theta_1$. In addition, assume there is no mass flux out of the boundaries, which using (13.33)$_3$ is equivalent to

$$\mathbf{k} \cdot \lambda_{\alpha\beta}\nabla c_\alpha = 0, \quad \text{on} \quad z = 0, d. \qquad (13.45)$$

Morro & Straughan (1990) choose to omit the $m_\alpha \mu_\alpha$ term in (13.41), since its presence requires a numerically determined base state. Instead they seek a steady solution of form

$$\bar{\theta} = -\gamma z + \theta_1, \qquad \bar{\mathbf{v}} = \mathbf{0}, \qquad (13.46)$$

where $\gamma = (\theta_1 - \theta_2)/d$ is the temperature gradient. The steady concentrations $\bar{c}_\alpha(z)$ satisfy the equations

$$\lambda_{\alpha\beta} \bar{c}_\beta'' - \tau_{\alpha\beta} c_\beta = \delta_\alpha \bar{\theta}(z), \qquad (13.47)$$

where a prime indicates differentiation with respect to the variable z. In Morro & Straughan (1990) investigation is confined to the case of one constituent, $c = c_1$: this has applications to a dissociating gas.

For simplicity put $\lambda = \lambda_{11}$, $\tau = \tau_{11}$ $\delta = \delta_1$, $\beta = \beta_1$, $\epsilon = \epsilon_1$. Equation (13.47) becomes for $\bar{c}(z)$

$$\lambda \bar{c}'' - \tau \bar{c} = \delta(\theta_1 - \gamma z),$$

and the boundary conditions (13.45) are

$$\bar{c}' = 0, \qquad z = 0, d. \qquad (13.48)$$

The solution is found to be

$$\bar{c} = \frac{\delta\beta}{\tau} \left\{ -\frac{1}{\epsilon} \sinh \epsilon z + \cosh(\epsilon d - 1) \cosh \epsilon z + z \right\} - \frac{\delta\theta_1}{\tau}, \qquad (13.49)$$

where $\epsilon = (\tau/\lambda)^{1/2}$. It is noteworthy to compare (13.49) with the non-reacting case which has $\bar{c}$ constant: here the nonlinear dependence on z is a direct consequence of the reaction.

A linear instability analysis of the steady solution (13.46), (13.49) commences with the introduction of perturbations $\mathbf{u}, \vartheta, \phi$, and π to $\bar{\mathbf{v}}, \bar{\theta}, \bar{c}$, and the steady pressure $\bar{p}$. From (13.41)–(13.43) the perturbations are found to satisfy

$$\vartheta_{,t} - \frac{\epsilon}{\bar{c}} \bar{\theta} \phi_{,t} = -\bar{\theta}' w + \frac{\epsilon}{\bar{c}} \bar{\theta} \bar{c}' w + K\Delta\vartheta, \qquad (13.50)$$

$$\rho\phi_{,t} = -\rho w \bar{c}' + \lambda \Delta\phi - \tau\phi - \delta\vartheta, \qquad (13.51)$$

$$\mathbf{u}_{,t} = -\frac{1}{\rho} \nabla\pi + \nu\Delta\mathbf{u} + \alpha g\vartheta\mathbf{k} - g\beta\phi\mathbf{k}, \qquad (13.52)$$

where $K = \kappa/\rho\tilde{c}$, $w = u_3$, and g is the acceleration due to gravity.

If one restricts attention to stationary convection and employs normal modes in (13.50)–(13.52) in such a way that the x, y dependence of $\mathbf{u}, \vartheta, \phi, \pi$ has a periodic shape with wavenumber ε, then after elimination of $\vartheta, \phi, u_1, u_2$, and π one may produce the following non-dimensional equation which w must satisfy on $z \in (0, 1)$,

$$
\Delta^4 w - R(1 - fS\Upsilon)\Delta\Delta^* w + \Lambda R(1 - fS\Upsilon)\Delta^* w \\
- \Lambda\Delta^3 w + R\Upsilon A\Delta(\mathcal{F}w) - \Upsilon R(B - fC)\varepsilon^2 w = 0,
\tag{13.53}
$$

where Λ, Υ are concentration and thermal reaction terms, respectively, and

$$
\Delta^* \equiv \partial^2/\partial x^2 + \partial^2/\partial y^2, \qquad R = \frac{\alpha\gamma g d^4}{K\nu} \quad \text{(Rayleigh number)},
$$

$$
\Lambda = \frac{\tau d^2}{\lambda}, \qquad \Upsilon = \frac{\gamma d\delta}{\tau}, \qquad S = \frac{\epsilon}{\tilde{c}} \qquad A = \frac{\beta K\rho}{\alpha\lambda\gamma d},
$$

$$
B = \frac{\beta d\tau}{\alpha\gamma\lambda}, \qquad C = \frac{d^2\rho\epsilon K\beta}{\lambda\alpha}.
$$

The function $\mathcal{F}$, which is the non-dimensional z-part of $\tilde{c}'$, is defined by

$$
\mathcal{F} = 1 - \sinh\Lambda^{1/2} z + \cosh(\Lambda^{1/2} - 1)\cosh\Lambda^{1/2} z
$$

and in (13.53) $f = z\mathcal{F}$.

Morro & Straughan (1990) present an asymptotic solution for the approximate free boundary conditions

$$
\frac{d^{2n}W}{dz^{2n}} = 0, \quad z = 0, 1; \qquad n = 0, 1, \dots,
$$

where $W(z)$ is the z-part of w. This allows one to see the first order effects of the reaction term Υ and the concentration reaction term Λ. Indeed the asymptotic analysis is probably valuable, since the reaction rates of many atmospheric reactions are typically of order 10^{-12}, see e.g. Kaye & Rood (1989). By writing

$$
W = W_{00} + \Upsilon W_{10} + \Lambda W_{01} + \cdots
$$

and

$$
R = R_{00} + \Upsilon R_{10} + \Lambda R_{01} + \cdots,
$$

Morro & Straughan (1990) find

$$
R = \frac{27\pi^4}{4}\left\{1 + \Upsilon\left(\frac{S}{4} - \frac{A}{\pi^2} - \frac{B}{3\pi^2}\right) + O(\Upsilon^2, \Upsilon\Lambda, \Lambda^2)\right\},
$$

which shows the thermal reaction term may have an important effect.

We conclude this chapter by pointing out that studies of multicomponent mixtures with or without reactions are likely to be of much practical value and are certainly mathematically very rich, as may be seen from Pearlstein et al. (1989), Terrones & Pearlstein (1989), and the references therein.

Of particular interest in the context of the current book is the problem of nonlinear energy stability for convection in a non-reacting fluid when there is a temperature field and two or more solutes present. If for example, there are two different solutes then a study of the case where the layer is heated below and heavier below in one constituent but top heavy in the other is not a symmetric problem and a generalized energy analysis should be very rewarding. The work of Joseph (1970) could well be very useful here, especially if a nonlinear unconditional result can be found. Another approach to this problem will be to use the interesting prescription for finding a generalized energy given by Galdi & Padula (1990); their theory of weakly coupled systems being relevant.

Another class of open problems are those involving convection for the full system (13.41–(13.43). Here it would appear that the basic state will have to be numerically determined and a stability analysis proceed from this. Despite this mathematical complication a stability analysis should prove worthwhile in view of the many applications.

14
Energy Stability and Other Continuum Theories

A large part of this book has been concerned with an up to date account of developments in the application of generalized energy methods in nonlinear convection problems. If any relevant literature has been omitted this is entirely unintentional. In this concluding chapter we give a brief indication of some areas where other generalized energy techniques are finding novel applications and we also briefly explore possible future uses of the energy method.

The energy-Casimir (convexity) method is an interesting one, which is appropriate to the investigation of the stability of inviscid flows and the stability of water waves (among other things). It would appear that this method is attributed to Arnold (1965a,b,1966a,b) and Holm et al. (1985) provides an extensive account. This is a stimulating paper and draws distinct attention to differences in the ideas of formal stability and true nonlinear stability. Due to the numerous and diverse applications of these ideas to e.g., magnetohydrodynamics, Bernstein et al. (1958), rotating neutron stars, Roberts (1981), and the connection with the energy integral method, e.g., Lutyen (1987), and the virial method of rotating fluid bodies, Chandrasekhar (1987), the energy-Casimir method is undoubtedly an important one. The paper by Holm et al. (1985) describes several applications and gives many references to work involved with this technique. The recent review of the energy-momemtum method of Simo et al. (1990) is also valuable in this context. This paper develops and applies the energy-momentum method to the problem of nonlinear stability of relative equilibria. Although the report concentrates on detailed applications to the stability of uniformly rotating states of models for geometrically exact rods and to a rigid body with an attached flexible appendage, it is relevant to mention it here. The energy-momentum technique of Simo et al. (1990) introduces a block diagonalization procedure, in which the second variation of the energy augmented with the linear and angular momentum block diagonalizes, separating the rotational from the internal vibration modes. Further references to the extensive applications of these ideas may be found in Simo et al. (1990).

The dynamo problem is an area where energy methods can be fruitful. While I am unaware of any applications of energy theory to the thermal–convection–dynamo problem, there have been several "anti-dynamo" theorems proved by variants of energy techniques. A very nice review of this topic is given by Fearn et al. (1988), pp. 110-130. In particular, L_∞ estimates have been the subject of recent investigations, cf. Lortz & Meyer-Spasche (1982), Ivers & James (1984), Stredulinsky et al. (1986); see also Ivers & James (1986).

Other recent work in the area of energy stability has been concerned with developing a theory for constructing, *on a systematic basis*, a suitable *generalized energy*, rather than proceeding on an ad hoc basis, as was done for example, for the rotating Bénard problem in chapter 6. Galdi & Padula (1990) is a notable contribution to this subject: they show how an appropriate generalized energy should involve combinations of the skew-symmertric part of the linear operator. Rionero (1988), Rionero & Mulone (1988), and Mulone & Rionero (1989), also give physical reasons why they believe certain combinations of terms should be present in a generalized energy; the relevant combinations they call essential variables.

In fact, the new energy theory of Galdi & Padula (1990), and in particular their use of energy analysis to provide an *instability principle*, see also Padula (1988), has substantial potential. On the stability side, the paper of Galdi & Padula (1990) is largely concerned with conditional stability. For those problems for which a coupling method yields good unconditional results, the extra complication introduced by their theory is unnecessary; however, their technique does yield very sharp stability boundaries to several problems for which conservative boundaries are achieved using coupling parameter and related methods.

Further work is ongoing in patterned ground formation. For example, the formation of polygonal ground structures underwater, as is found in shallow lakes in Wyoming, is under theoretical investigation. Convection problems involving temperature dependent viscosity are also proving amenable to energy methods by appropriate use of the Sobolev inequality. The latter class of problem, in particular, has necessitated another type of generalized energy.

Yet another difficult problem currently being tackled by energy theory is that of compressible convection, see Spiegel (1965). Padula (1986) was the first to address this by a generalized energy method, and later Coscia & Padula (1990). The energy functional deemed necessary is non-trivial. I am unaware of numerical results for the energy stability threshold.

As is observed by Lindsay & Straughan (1990), two areas of continuum mechanics undergoing much activity are *mixture theories* and *extended thermodynamics*. The latter subject has been developed extensively by Professor I. Müller and his co-workers, cf. Müller (1985), Müller & Ruggeri (1987), Jou et al. (1988). It would appear that little stability work has been undertaken in this area, although it is a fruitful area of study, particularly by the energy method. However, the systems are usually hyperbolic and so methods different from those discussed here will be necessary.

The novel applications of mixture theories to concrete, mundane problems are very promising. As examples we cite the work on phase boundaries between ice and water with applications in the physics of temperate glaciers by Alts & Hutter (1988a,b,c,1989), Hutter (1982,1983), Hutter & Engelhardt (1988), and Hutter et al. (1988), in the modelling of snow movement by Morland et al. (1990), in the modelling of mushy zones and slurries by Hills & Roberts (1987a,b,1988a,b), in the modelling of compositional convection at the Earth's core by Loper & Roberts (1978,1980), and on the modelling of chemically reacting fluid mixtures, cf. chapter 13. In these theories there are many challenging stability problems. Their nature is very much one of a compressible system and even a linearized instability analysis will be both useful and difficult. If an energy theory can be developed for such problems, the results should be worthwhile.

Appendix 1

Some Useful Inequalities in Energy Stability Theory

The purpose of this appendix is simply to collect some of the inequalities that have been found to be very useful in energy stability theory. The inequalities presented are not necessarily the most general forms available, but are given in the form in which I have seen them used.

The Poincaré Inequality. Let V be a "cell" in three dimensions. Suppose for simplicity V is the cell $0 \le x < 2a_1$, $0 \le y < 2a_2$, $0 < z < 1$, and suppose u is a function periodic in x, y, of period $2a_1, 2a_2$, respectively, and $u = 0$ on $z = 0, 1$. Then the Poincaré inequality may be written

$$< u^2 > \le \frac{1}{\pi^2} < u_{i,j} u_{i,j} >, \tag{1}$$

where $< \cdot >$ denotes integration over V. In general, the constant, $1/\pi^2$ in (1), depends on the geometry and size of the domain V.

The Wirtinger Inequality. Suppose the boundary conditions on u above are replaced by

$$\frac{\partial u}{\partial z} = 0, \qquad \text{on} \qquad z = 0, 1. \tag{2}$$

For functions such that

$$< u > = 0, \tag{3}$$

u is periodic in x, y with periods $2a_1, 2a_2$; the Wirtinger inequality is

$$< u^2 > \le \frac{1}{\pi^2} < u_{i,j} u_{i,j} > . \tag{4}$$

To establish (4) we seek the maximum

$$\lambda_1 = \max_{\mathcal{H}} \frac{\|f\|^2}{\|\nabla f\|^2}, \tag{5}$$

where

$$\mathcal{H} = \Big\{ u \in C^2(V) \cap C^1(\bar{V}) \Big| u \text{ periodic in } x, y,$$
$$\frac{\partial u}{\partial z} = 0 \quad \text{on} \quad z = 0, 1, \quad \text{and} \quad < u > = 0 \Big\}. \tag{6}$$

(When no boundary conditions are specified, but the condition that $< u > = 0$ is still required, Hardy et al. (1934), p. 184, show that $\lambda_1 = 4/\pi^2$, for functions of one variable.) To resolve (5) we take the variation to find

$$< f\psi + \lambda_1 \psi \Delta f + \lambda_2 \psi > = 0,$$

for all $\psi \in C^1(V) \cap C(\bar{V})$, with λ_2 being a Lagrange multiplier. This leads to

$$f + \lambda_1 \Delta f = -\lambda_2.$$

We integrate this over V and use the boundary conditions on f to see that $\lambda_2 = 0$. A normal mode expansion is taken for f, i.e.,

$$f = F(z)e^{i(mx+ny)}, \quad \text{where} \quad F = \sum_{r=1}^{\infty} A_r \cos r\pi z.$$

This leads to

$$\lambda_1 = \max \frac{1}{r^2\pi^2 + a^2}, \quad \text{where} \quad a^2 = m^2 + n^2.$$

The maximum is achieved for $r = 1$ and $a^2 \to 0$, and (4) follows.

The Sobolev Inequality. The general Sobolev embedding inequality may be found in, e.g., Gilbarg & Trudinger (1977), pp. 148–157. The one of frequent use in energy stability theory is the following. Let Ω be a bounded domain in $\mathbf{R}^3$ with boundary $\partial\Omega$. Then for functions u with $u = 0$ on $\partial\Omega$,

$$\left(\int_\Omega u^6 \, dV \right)^{1/3} \leq C \int_\Omega |\nabla u|^2 \, dV, \tag{7}$$

where the constant C is independent of the domain; in fact,

$$C^3 = \frac{4}{3^{3/2}\pi^2}.$$

An Inequality for the Supremum of a Function. We calculate a value for the constant C in the inequality

$$\sup_V |u(\mathbf{x})| \leq C\|\Delta u\|, \tag{8}$$

where V is the cell $V = [0, 2a_1) \times [0, 2a_2) \times (0, 1)$, and $\mathcal{H}$ is given by (6). Inequality (8) is established as in Galdi & Straughan (1985b).

Let $\mathbf{x}$ be a point in V, and let $\mathcal{C}$ be a cone, vertex $\mathbf{x}$, whose base is part of the sphere radius h, cf. Adams (1975), p. 107. Then,

$$u(\mathbf{x}) \equiv u(0, \theta) = u(r, \theta) - \int_0^r \frac{d}{dt} u(t, \theta) \, dt.$$

Hence, for all $r \in (0, h]$,

$$u(\mathbf{x}) \equiv u(0, \theta) \leq u(r, \theta) + \int_0^h |\text{grad}\, u(t, \theta)| \, dt. \tag{9}$$

Denote by $\omega(\theta)$ the solid angle in $\mathcal{C}$, multiply (9) by $\omega(\theta) r^2$ and integrate over $\mathcal{C}$ to find, with $\Phi =$ volume of $\mathcal{C}$,

$$\Phi|u(\mathbf{x})| \leq \int_{\mathcal{C}} |u(r, \theta)| dV + \int_0^h r^2 \, dr \int_{\mathcal{C}} \frac{|\text{grad}\, u|}{r^2} \, dV, \tag{10}$$

where dV is the volume element. By Hölder's inequality it may be deduced from (10) that

$$\Phi|u(\mathbf{x})| \leq \Phi^{1/2} \left(\int_{\mathcal{C}} |u|^2 dV \right)^{1/2}$$
$$+ \frac{h^3}{3} \left(\int_{\mathcal{C}} |\text{grad}\, u|^6 \, dV \right)^{1/6} \left(\int_{\mathcal{C}} \frac{\sin\phi}{r^{2/5}} \, d\phi \, d\theta \, dr \right)^{6/5}. \tag{11}$$

The quantity Φ and the last integral in (11) are evaluated, and (11) is extended to the cell V to deduce that for the angle in $\mathcal{C}$ to be $\frac{1}{2}\pi$ and

$$h = \min\{a_1, a_2, 1\}, \tag{12}$$

$$\Phi|u(\mathbf{x})| \leq \Phi^{1/2}\|u\| + \frac{h^3}{3} I \left(\int_V |\text{grad}\, u|^6 dV \right)^{1/6}, \tag{13}$$

where I and Φ are given by

$$\Phi = \frac{1}{3}\pi h^3 2^{1/2}(2^{1/2} - 1), \qquad I = \frac{5}{3}\pi 2^{1/2}(2^{1/2} - 1)h^{3/5}.$$

The next step is to estimate the last term in (13). To begin we integrate by parts and then use the Cauchy-Schwarz inequality to find

$$< u_{,i} u_{,i} > = - < u \Delta u >$$
$$\leq \|u\| \, \|\Delta u\|. \tag{14}$$

We next use the Wirtinger inequality, (4), in (14) to find

$$< u_{,i}u_{,i} > \leq \pi^{-2}\|\Delta u\|^2. \tag{15}$$

For a cell such as V the Sobolev inequality, cf. Adams (1975), p. 104, may be written

$$\|w\|_{L^6} \leq 2^{5/2}\|w\|_{H^1}, \tag{16}$$

where $\| \cdot \|_{L^6}$ and $\| \cdot \|_{H^1}$ are the norms on $L^6(V)$ and $H^1(V)$; so with $w = \nabla u$,

$$\|\nabla u\|_6 \leq 2^{5/2}\left(\|\nabla u\|^2 + < u_{,ij}u_{,ij} >\right)^{1/2}. \tag{17}$$

To deal with the last term in (17) we integrate by parts to find

$$< u_{,ij}u_{,ij} > = - < u_{,j}\Delta u_{,j} > + \int_{\partial V} n_i u_{,j}u_{,ji}\, dA$$
$$= \|\Delta u\|^2 + \int_{\partial V} (n_i u_{,j}u_{,ji} - n_i u_{,i}\Delta u)dA.$$

The boundary conditions on u ensure the boundary integral is zero. Hence from (17),

$$\|\nabla u\|_6 \leq 2^{5/2}\left(\|\nabla u\|^2 + \|\Delta u\|^2\right)^{1/2}$$
$$\leq \frac{2^{5/2}}{\pi}\sqrt{1 + \pi^2}\,\|\Delta u\|, \tag{18}$$

where in the last line we have used (15).

Inequality (18) is inserted into (13) and the Wirtinger inequality (4) is used on the $\|u\|$ term in (13) to arrive at

$$\Phi|u(\mathbf{x})| \leq \left(\frac{\Phi^{1/2}}{\pi^2} + \frac{2^{5/2}h^3}{3\pi}(1 + \pi^2)^{1/2}\,I\right)\|\Delta u\|. \tag{19}$$

Inequality (19) holds $\forall \mathbf{x} \in V$ and so we divide by Φ and replace $|u(\mathbf{x})|$ by $\sup |u(\mathbf{x})|$. This establishes (18) with a value for C of

$$C = \frac{\sqrt{3}}{\left[\pi^5 h^3\sqrt{2}(2^{1/2} - 1)\right]^{1/2}} + \frac{2^{5/2}5(1 + \pi^2)^{1/2}h^{3/5}}{3\pi}. \tag{20}$$

A Sobolev Inequality for u^4. For a cell V we employ Hölder's inequality to deduce for functions u with x periodicity a_1, y periodicity a_2,

$$\int_V u^4\, dV \leq (a_1 a_2)^{1/3}\left(\int_V u^6\, dV\right)^{2/3}.$$

If we combine this with the Sobolev inequality (16) we find

$$< u^4 >^{1/2} \le 32(a_1 a_2)^{1/6} \big(< u^2 > + < u_{,i} u_{,i} > \big). \tag{21}$$

This is a general inequality that does *not* require boundary conditions to be specified on u at $z = 0, 1$.

We now present a version of (21) for a general cell shape.

A Sobolev Inequality for u^4 over a More General Cell. We first present a general inequality, cf. Galdi et al. (1987). For simplicity we assume:

(i) Λ is a cross section of a cell in the layer $\{z \in (0,1)\}$ and is star shaped with respect to an origin in Λ;

(ii) lines parallel to the x and y axes intersect $\partial\Lambda$ in at most two points.

The configuration is shown in Figure A1.1.

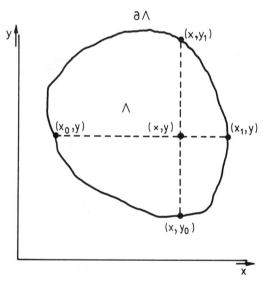

Figure A1.1. Sketch of the cross section of a cell.
(After Galdi et al. (1987).)

Let u be any function that is zero on $z = 0$. Then

$$u^2(x, y, z) = u^2(x_1, y, z) - 2 \int_x^{x_1} u(s, y, z) \frac{\partial u}{\partial s}(s, y, z) ds,$$

$$u^2(x, y, z) = u^2(x_0, y, z) + 2 \int_{x_0}^x u(s, y, z) \frac{\partial u}{\partial s}(s, y, z) ds,$$

from which we find

$$u^2(x, y, z) \le \frac{1}{2} \left[u^2(x_0, y, z) + u^2(x_1, y, z) + 2 \int_{x_0}^{x_1} |u|\, |u_s| ds \right].$$

In a similar manner we may show

$$u^2(x, y, z) \le \frac{1}{2} \Big[u^2(x, y_0, z) + u^2(x, y_1, z) + 2 \int_{y_0}^{y_1} |u| \, |u_t| dt \Big].$$

These inequalities are multiplied, and we integrate to obtain

$$\int_\Lambda u^4 \, dA \le \Big[\frac{1}{2} \int_{\partial\Lambda} u^2 |n_1| ds + \int_\Lambda |u| \, |u_x| \, dA \Big] \qquad (22)$$
$$\times \Big[\frac{1}{2} \int_{\partial\Lambda} u^2 |n_2| ds + \int_\Lambda |u| \, |u_y| \, dA \Big],$$

where n_i are the components of the unit outward normal to $\partial\Lambda$ and ds denotes integration along $\partial\Lambda$. We next employ the arithmetic-geometric mean inequality in (22), and then observe that

$$\int_{\partial\Lambda} u^2 |n_1| ds \int_{\partial\Lambda} u^2 |n_2| ds \le \frac{1}{2} \Big(\int_{\partial\Lambda} u^2 ds \Big)^2,$$

to derive from (22)

$$\int_\Lambda u^4 \, dA \le \frac{1}{4} \Big(\int_{\partial\Lambda} u^2 ds \Big)^2 + \int_\Lambda u^2 \, dA \int_\Lambda u_{,\alpha} u_{,\alpha} \, dA, \qquad (23)$$

where α sums over 1 and 2. Furthermore,

$$\int_{\partial\Lambda} u^2 ds \le \frac{1}{\sigma} \int_{\partial\Lambda} x_\alpha n_\alpha u^2 ds = \frac{2}{\sigma} \int_\Lambda u^2 \, dA + \frac{2}{\sigma} \int_\Lambda x_\alpha u u_{,\alpha} \, dA, \qquad (24)$$

in which

$$\sigma = \min_{\partial\Lambda} x_\alpha n_\alpha.$$

Let

$$M = \max_\Lambda |x_\alpha|,$$

and then from (24) we may derive

$$\int_{\partial\Lambda} u^2 ds \le \frac{2}{\sigma} \int_\Lambda u^2 \, dA + \frac{2M}{\sigma} \Big[\int_\Lambda u^2 \, dA \int_\Lambda u_{,\alpha} u_{,\alpha} \, dA \Big]^{1/2}. \qquad (25)$$

We put (25) in (23) and then integrate with respect to z to obtain

$$< u^4 > \; \le \; \max_z \Big(\int_\Lambda u^2 \, dA \Big)$$
$$\times \Big[\frac{1}{\sigma^2} \|u\|^2 + \Big\{ 1 + \Big(\frac{M}{\sigma} \Big)^2 \Big\} \|\nabla^* u\|^2 + \frac{2M}{\sigma^2} \|u\| \, \|\nabla^* u\| \Big], \qquad (26)$$

where $\nabla^* \equiv (\partial/\partial x, \partial/\partial y)$ and $\|\cdot\|$ dentoes the L^2 norm on the cell V. To estimate the Λ term in (26) we note

$$\max_z \left(\int_\Lambda u^2 \, dA \right) \leq 2\|u\| \, \|u_z\|. \tag{27}$$

Also, for the boundary conditions on u, the Wirtinger inequality is

$$\|u\|^2 \leq \frac{4}{\pi^2} \|u_z\|^2. \tag{28}$$

Then, use of (27), (28) in (26) provides

$$< u^4 > \leq \gamma^4 \|\nabla u\|^2, \tag{29}$$

where

$$\gamma^4 = \frac{16}{\sigma^2 \pi^3} \left[1 + M\pi + \frac{1}{4}\pi^2(\sigma^2 + M^2) \right]. \tag{30}$$

It should be noted that the above derivation does not assume periodicity in x and y. Furthermore, when $u = 0$ on both $z = 0, 1$, inequality (29) holds but with the constant in (30) sharpened to

$$\gamma^4 = \frac{1 + 2M\pi + \pi^2(\sigma^2 + M^2)}{\sigma^2 \pi^3}.$$

When the periodicity conditions are taken into account we may further sharpen (26) and (29). For a hexagon with side length $4\pi/3a$ and solution wavelength a, (26), (29) continue to hold but the constants may be replaced by

$$\gamma^4 = \frac{36a^2 + 48a\pi^2 + 28\pi^4}{3\pi^5} \quad \text{or} \quad = \frac{9a^2 + 24a\pi^2 + 28\pi^4}{12\pi^5}, \tag{30'}$$

respectively.

A Two-Dimensional Surface Inequality. The following two inequalities and the counterexample were established in Galdi et al. (1987).

We establish a value for c in the *two-dimensional* inequality, namely,

$$\int_\Gamma u^3 dA \leq c\|u\|D(u), \tag{31}$$

for $u = 0$ at $z = 0$, and where Γ is the region $x \in (0, k)$ and V the two-dimensional cell $\Gamma \times (0, 1)$. $\|\cdot\|$ denotes the L^2 norm on V and $D(\cdot)$ the Dirichlet integral there.

The procedure begins by noting

$$\int_\Gamma u^3 dA = 3 \int_V u^2 \frac{\partial u}{\partial z} \, dV.$$

The Cauchy-Schwarz inequality is applied to obtain

$$\int_\Gamma u^3 dA \le 3D^{1/2}(u)\left(\int_V u^4 \, dV\right)^{1/2}. \tag{32}$$

The u^4 term is handled by Joseph's (1976a) method. Since $u(x,0) = 0$ and u is $x-$periodic of period k, it follows that for $x \in (\hat{x}, \hat{x} + k)$, with $\hat{x}$ arbitrary but fixed,

$$u^2(x,z) = 2\int_0^z u(x,t)\frac{\partial u}{\partial t}(x,t)dt,$$

$$u^2(x,z) \le \int_{\hat{x}}^{\hat{x}+k} |u(s,z)|\left|\frac{\partial u}{\partial s}\right|ds + u^2(\hat{x},z).$$

We multiply these expressions and integrate over $(\hat{x}, \hat{x} + k) \times (0,1)$ to find

$$\int_0^1 \int_{\hat{x}}^{\hat{x}+k} u^4(x,z)\,dx\,dz \le 2\int_0^1\int_{\hat{x}}^{\hat{x}+k}|u|\,|u_z|\,dx\,dz \int_0^1\int_{\hat{x}}^{\hat{x}+k}|u|\,|u_x|\,dx\,dz$$
$$+ 2\int_0^1 u^2(\hat{x},z)\,dz \int_0^1\int_{\hat{x}}^{\hat{x}+k}|u|\,|u_x|\,dx\,dz.$$

Only the single integral in the expression above depends on $\hat{x}$, since u is a periodic function in x, of period k. Therefore, we integrate this inequality over $(\hat{x}, \hat{x} + k)$ and set $\hat{x} = 0$, and then use the Cauchy-Schwarz inequality to find

$$\int_0^1 \int_0^k u^4(x,z)\,dx\,dz \le 2\|u\|^2\,\|u_x\|\,\|u_z\| + \frac{2}{k}\|u\|^3\,\|u_x\|.$$

Finally, we use Poincaré's inequality (for a function that vanishes only at $z = 0$) to deduce

$$\int_0^1 \int_0^k u^4(x,z)\,dx\,dz \le 2\|u\|^2\,\|\nabla u\|^2 + \frac{4}{\pi k}\|u\|^2\,\|\nabla u\|^2. \tag{33}$$

Inequality (31) follows by using (33) in (32), with a value of c given by

$$c = 3\sqrt{2 + \frac{4}{\pi k}}. \tag{34}$$

Inequality (31) *is False in Three-Dimensions.* To provide a counterexample to (31) in three dimensions consider the region

$$V = (0,2) \times (0,2) \times (0,1).$$

Define a C^∞ cut-off function by

$$h = \begin{cases} 1, & r \equiv |\mathbf{x} - P| \le \frac{1}{2}, \\ \exp\left[\frac{4}{3} + (r^2 - 1)^{-1}\right], & \frac{1}{2} \le r \le 1, \\ 0, & r \ge 1, \end{cases}$$

where $P = (1,1,1)$, on the cell V. For $\alpha > 1$, a constant to be specified, define f by

$$f(\mathbf{x}) = h \exp(-\alpha r),$$

and define the quantity F by

$$F = \frac{\int_\Gamma |f|^3 dA}{\|f\| \, \|\nabla f\|^2},$$

where Γ is the $z = 1$ boundary of V. Since $|h|, |h'|$ are bounded and $h \equiv 1$ for $r \in (0, \frac{1}{2})$, we put $\rho^2 = (x - 1)^2 + (y - 1)^2$ to deduce

$$F \ge B \frac{\int_0^{\frac{1}{2}} \exp(-3\alpha\rho)\, \rho\, d\rho}{\alpha^2} \left(\int_0^1 \exp(-2\alpha r)\, r^2\, dr \right)^{3/2}.$$

In this expression B is a positive constant, independent of α. The change of variables $u = \alpha\rho$, $v = \alpha r$, leads to

$$F \ge \text{constant} \times \sqrt{\alpha}, \qquad \alpha \to \infty.$$

This establishes (32) is not true in three dimensions.

A Boundary Estimate for u^2. Let now V be a three-dimensional periodicity cell contained in $z \in (0,1)$. Let also Γ be that part of the boundary of V that intersects $z = 1$.

We establish the following *isoperimetric* inequality for functions u that vanish on $z = 0$,

$$\int_\Gamma u^2\, dA \le 2\|u\| \, \|\nabla u\|. \tag{35}$$

This follows since

$$u^2(x, y, 1) = 2 \int_0^1 u \frac{\partial u}{\partial z}\, dz;$$

integration over V and use of the Cauchy-Schwarz inequality completes the derivation. To see inequality (35) is the best possible, choose $u = \sinh \alpha z$, then

$$\frac{\|u\| \, \|\nabla u\|}{\int_\Gamma u^2\, dA} \to \frac{1}{2},$$

for $\alpha \to \infty$.

A Surface Inequality for u^4. Suppose $u(x,y)$ is periodic in x, y of x−period k and of y−period m. Define Γ to be the rectangle $(\hat{x}, \hat{x} + k) \times (\hat{y}, \hat{y} + m)$, $\hat{x}$, $\hat{y}$ fixed. We now show that provided

$$\int_\Gamma u\,dA = 0,$$

$$\int_\Gamma u^4\,dA \le \left[1 + \frac{4}{\pi^2} + \frac{2(k+m)}{\pi\sqrt{km}}\right]\int_\Gamma u^2\,dA \int_\Gamma u_{;\alpha}u_{;\alpha}dA. \qquad (36)$$

The proof employs Joseph's method (1976a), p. 249, and was given by Straughan (1991). Since u has x−period k, y−period m, we write

$$2\int_{\hat{x}}^x u(s,y)u_s(s,y)\,ds + u^2(\hat{x}, y) = u^2(x,y)$$

$$= -2\int_x^{\hat{x}+k} u(s,y)u_s(s,y)\,ds + u^2(\hat{x}, y)$$

and

$$2\int_{\hat{y}}^y u(x,t)u_t(x,t)\,dt + u^2(x, \hat{y}) = u^2(x,y)$$

$$= -2\int_y^{\hat{y}+m} u(x,t)u_t(x,t)\,dt + u^2(x, \hat{y}).$$

From these expressions it is easily seen that

$$u^2(x,y) \le \int_{\hat{x}}^{\hat{x}+k} |u(s,y)|\,|u_s(s,y)|\,ds + u^2(\hat{x}, y), \qquad (37)$$

$$u^2(x,y) \le \int_{\hat{y}}^{\hat{y}+m} |u(x,t)|\,|u_t(x,t)|\,dt + u^2(x, \hat{y}). \qquad (38)$$

We now multiply (37) and (38) together and integrate over Γ twice to find, with the help of the Cauchy-Schwarz inequality,

$$km\int_\Gamma u^4\,dA \le km\int_\Gamma u^2\,dA \int_\Gamma u_{;\alpha}u_{;\alpha}\,dA + \left(\int_\Gamma u^2\,dA\right)^2$$
$$+ (k+m)\left(\int_\Gamma u^2\,dA\right)^{3/2}\left(\int_\Gamma u_{;\alpha}u_{;\alpha}\,dA\right)^{1/2}. \qquad (39)$$

The Wirtinger inequality, in two dimensions, shows

$$\int_\Gamma u^2\,dA \le \frac{4km}{\pi^2}\int_\Gamma u_{;\alpha}u_{;\alpha}\,dA.$$

We put this into (39) and divide by km to obtain (36).

Appendix 2

Numerical Solution of the Energy Eigenvalue Problem

A2.1 The Shooting Method

There are many detailed descriptions available in the literature of methods for finding eigenvalues and eigenfunctions for partial differential equations. This appendix is here only for completeness. Its purpose is to describe a method for solving eigenvalue problems of the type encountered in linear and energy stability convection problems. The technique referred to is the compound matrix method, which is simple to implement, but which is nevertheless very accurate and efficient. First we briefly describe a standard shooting method.

We begin at an elementary level and investigate the second order eigenvalue problem,

$$\frac{d^2u}{dx^2} + \lambda u = 0, \qquad 0 < x < 1,$$

$$u(0) = u(1) = 0. \tag{1}$$

This problem arises in chapter 2, where it was seen that

$$\lambda_n = n^2\pi^2. \tag{2}$$

For stability studies it is usually only the smallest eigenvalue that is of interest, i.e., λ_1.

To determine λ_1 numerically we retain $u(0) = 0$, but replace $u(1) = 0$ by a prescribed condition on $u'(0)$, a convenient choice being $u'(0) = 1$, thereby converting the *boundary value problem* to an *initial value* one. Two values of λ_1 are selected, say $0 < \lambda_1^{(1)} < \lambda_1^{(2)}$, and then the initial value problem is integrated numerically to find $u_1(1)$, $u_2(1)$, where u_i denotes the solution corresponding to $\lambda_1^{(i)}$: (a high order, variable step Runge-Kutta-Verner technique is often adequate for the numerical integration). The idea is to use $u_1(1)$, $u_2(1)$ so found to ensure $u(1)$ is as close to zero as required by some pre-specified degree of accuracy. An iteration technique is then employed to find a sequence $u_k(1)$, corresponding to $\lambda_1^{(k)}$, such that

$$|u_k(1)| < \epsilon, \qquad k \text{ large enough}, \tag{3}$$

where ϵ is a user specified tolerance. I have found the secant method, see e.g., Cheney & Kincaid (1985), p. 97, is a suitable routine for this purpose. Once a $u_k(1)$ is determined to satisfy (3), $\lambda_1^{(k)}$ is then the required numerical estimate of the first eigenvalue to (1).

For practical purposes it is usually necessary to have some guide as to what values to select for $\lambda_1^{(1)}$, $\lambda_1^{(2)}$. However, in energy theory one can usually use linear stability theory as a guide for this. For linear stability theory it is often possible to use known results from related problems that can be solved analytically.

A2.2 A System: The Viola Eigenvalue Problem

Of course, the eigenvalue problems encountered in convection are more complicated than that discussed in §A2.1, but the basic numerical method is the same. To illustrate how the shooting method works on a system we use an eigenvalue problem studied by Viola (1941); see also Fichera (1978). This is

$$\frac{d^2}{dx^2}\left[(1-\theta x)^3\frac{d^2u}{dx^2}\right] - \lambda(1-\theta x)u = 0, \quad x \in (0,1),$$

$$u = \frac{d^2u}{dx^2} = 0, \quad x = 0,1, \tag{4}$$

where the constant θ satisfies $0 \le \theta < 1$.

To solve (4) for the first (lowest) eigenvalue λ_1 does not appear possible analytically, in general, i.e., for $\theta \ne 0$. This problem is described in detail by Fichera (1978), pp. 41–43, who gives an exposition of the orthogonal invariants method for obtaining very accurate upper and lower bounds to λ_k, $k = 1, 2, \ldots$. To solve (4) by a shooting method we first write it as a system of four first order differential equations in the vector $\mathbf{u} = (u, u', u'', u''')$, where $u' = du/dx$, $u'' = d^2u/dx^2$, etc. The boundary conditions at $x = 1$ on u, u'' are replaced in turn by $u' = 1$, $u''' = 0$, and then $u' = 0$, $u''' = 1$, at $x = 0$. The two initial value problems thereby obtained are then integrated numerically. Let the solution so found be written as a linear combination of the two solutions so obtained, say $\mathbf{u} = \alpha\mathbf{v} + \beta\mathbf{w}$. Then, the correct boundary condition $u = u'' = 0$ at $x = 1$ is imposed, and this requires

$$\det\begin{pmatrix} v & v'' \\ w & w'' \end{pmatrix} = 0 \tag{5}$$

to hold at $x = 1$.

While the shooting method is easy to understand and implement, it suffers from a serious drawback. This is that one has numerically to locate the zero of a determinant; e.g., in (5), one has to locate

$$v(1)w''(1) - w(1)v''(1) = 0.$$

The two quantities $v(1)w''(1)$ and $w(1)v''(1)$ must, therefore, be very close although neither need be close to zero (and generally will not). One is thus faced with subtracting two nearly identical quantities, and this can lead to very large round off errors and significant error build up during the solution of a convection eigenvalue problem. There are many ways to overcome this: we describe only one. This is the compound matrix technique, which has distinct advantages for energy eigenvalue problems. Basically the idea is to remove the troublesome location of the zero of a determinant by converting to a system of ordinary differential equations in the determinants themselves. A lucid description of the compound matrix method and how one uses it to find eigenvalues and eigenfunctions, for the Orr-Sommerville problem, is given by Drazin & Reid (1981). In this appendix we simply show how it is used to find the lowest eigenvalue to two systems of interest.

A2.3 The Compound Matrix Method

It is the purpose of this section to present an accurate numerical calculation of eigenvalues to (4). For $0 \leq \theta < 0.9$ accurate results are evidently easily found by the standard shooting method described in §A2.2. However, for the case where $\theta \to 1^-$ for which (4) becomes singular, other methods must be employed. Some useful information may be gleaned in this case with the compound matrix technique.

   To solve (4) by the compound matrix method we let $\mathbf{U} = (u, u', u'', u''')^T$, and then suppose $\mathbf{U}_1$ and $\mathbf{U}_2$ are solutions to (4) with values at $x = 0$ of $(0, 1, 0, 0)^T$ and $(0, 0, 0, 1)^T$, respectively. A new six vector

$$\mathbf{Y} = (y_1, y_2, y_3, y_4, y_5, y_6)^T$$

is defined as the 2×2 minors of the 4×2 solution matrix whose first column is $\mathbf{U}_1$ and second $\mathbf{U}_2$. So,

$$
\begin{aligned}
y_1 &= u_1 u_2' - u_1' u_2, \\
y_2 &= u_1 u_2'' - u_1'' u_2, \\
y_3 &= u_1 u_2''' - u_1''' u_2, \\
y_4 &= u_1' u_2'' - u_1'' u_2', \\
y_5 &= u_1' u_2''' - u_1''' u_2', \\
y_6 &= u_1'' u_2''' - u_1''' u_2''.
\end{aligned}
\tag{6}
$$

By direct calculation from (4) the initial value problem for the y_i is found to be

$$y_1' = y_2,$$
$$y_2' = y_3 + y_4,$$
$$y_3' = y_5 + 6\frac{\theta}{M}y_3 - 6\left(\frac{\theta}{M}\right)^2 y_2,$$
$$y_4' = y_5,$$
$$y_5' = y_6 + 6\frac{\theta}{M}y_5 - 6\left(\frac{\theta}{M}\right)^2 y_4 - \frac{\lambda}{M^2}y_1,$$
$$y_6' = 6\frac{\theta}{M}y_6 - \frac{\lambda}{M^2}y_2,$$

(7)

where

$$M = 1 - \theta x.$$

From the initial conditions on $\mathbf{U}_1$ and $\mathbf{U}_2$ we see that system (7) is to be integrated numerically subject to the initial condition

$$y_5(0) = 1 \tag{8}$$

and the final condition

$$y_2(1) = 0. \tag{9}$$

Again the zero in (9) is located to a pre-assigned degree of accuracy.

The numerical computations we report were all done on the University of Wyoming's CDC Cyber 760. The numerical results are very accurate in comparison with the available bounds of Fichera (1978), p. 43. For $\theta = 0.5$, Fichera finds

$$50.71623063 \le \lambda_1 \le 50.71623066, \qquad 838.2089 \le \lambda_2 \le 838.2091;$$

we find

$$\lambda_1 = 50.71623064799, \qquad \lambda_2 = 838.2090471111.$$

For $\theta = 0$, the exact values are $\lambda_1 = \pi^4$, $\lambda_2 = 16\pi^4$; we find

$$\lambda_1 = 97.40909103339, \qquad \lambda_2 = 1558.545456538.$$

Table A2.1 gives what we believe are accurate values of λ_1 for θ between 0 and 0.9. Figure A2.1 displays these and heuristic values for $0.9 \le \theta \le 0.999$. We have also calculated λ_2 for other values of θ, e.g.,

$$\theta = 0.05, \qquad \lambda_2 = 1481.214398565,$$
$$\theta = 0.1 \ , \qquad \lambda_2 = 1405.075275206.$$

What is clearly an evident fact is that the asymptotic behaviour of λ as $\theta \to 1^-$ appears to be a very interesting problem. The orthogonal invariants method of Fichera (1978) clearly needs a careful analysis for $\theta \sim 1^-$, since terms like $\log(1 - \theta)$ are involved. The numerical results *indicate* that λ_1 may approach 0 as $\theta \to 1^-$. The behaviour of λ_1 as $\theta \to 1^-$ is an open and interesting problem of analysis.

λ_1	θ
97.40909	0
92.55916	0.05
87.75027	0.1
82.98231	0.15
78.25510	0.2
73.56833	0.25
68.92154	0.3
59.74500	0.4
50.71623	0.5
41.81693	0.6
33.00917	0.7
24.20406	0.8
19.73541	0.85
15.12813	0.9

Table A2.1. Numerical values of λ_1 for various θ.

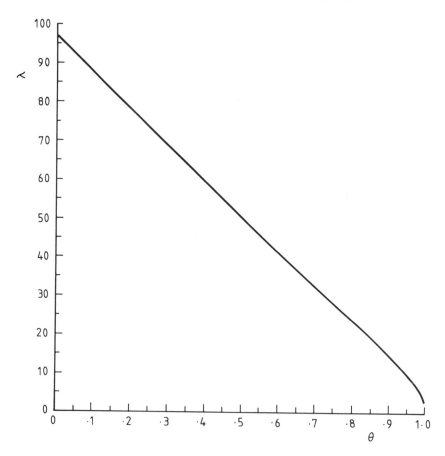

Figure A2.1. The behaviour of λ_1 against θ.

A2.4 Numerical Solution of (4.65), (4.66) Using Compound Matrices

To complete this appendix we give relevant details for the numerical solution of (4.65), (4.66) using the compound matrix method. We introduce the vector $\mathbf{W} = (W, W', \Theta, \Theta')^T$, and then let $\mathbf{W}_1$ and $\mathbf{W}_2$ be independent solutions to (4.65), (4.66) for initial values $\mathbf{W}_1(0) = (0, 1, 0, 0)^T$, $\mathbf{W}_2(0) = (0, 0, 0, 1)^T$, respectively. A new six vector $\mathbf{Y} = (y_1, ..., y_6)^T$ is formed from the 2×2 minors of the 4×2 solution matrix with first column

W_1 and second W_2. Thus,

$$
\begin{aligned}
y_1 &= W_1 W_2' - W_2 W_1', \\
y_2 &= W_1 \Theta_2 - W_2 \Theta_1, \\
y_3 &= W_1 \Theta_2' - W_2 \Theta_1', \\
y_4 &= W_1' \Theta_2 - W_2' \Theta_1, \\
y_5 &= W_1' \Theta_2' - W_2' \Theta_1', \\
y_6 &= \Theta_1 \Theta_2' - \Theta_2 \Theta_1'.
\end{aligned}
\tag{10}
$$

Using equations (4.65) we calculate the ordinary differential equations for the y_i to be

$$
\begin{aligned}
y_1' &= -\frac{1}{2} R_E F a^2 y_2, \\
y_2' &= y_3 + y_4, \\
y_3' &= a^2 y_2, \\
y_4' &= a^2 y_2, \\
y_5' &= a^2 (y_3 + y_4) + \frac{1}{2} R_E F y_1 - \frac{1}{2} R_E F a^2 y_6, \\
y_6' &= \frac{1}{2} R_E F y_2.
\end{aligned}
\tag{11}
$$

These equations are integrated numerically from 0 to 1. The boundary conditions on W are

$$
W(0) = W(1) = \Theta(0) = \Theta(1) = 0,
\tag{12}
$$

and we have replaced the conditions at $z = 1$ by

$$
W_1'(0) = 1, \qquad \Theta_2'(0) = 1.
\tag{13}
$$

It is not difficult to see from (10), (13), that the initial condition for (11) is

$$
y_5(0) = 1,
\tag{14}
$$

the other $y_i(0)$ being 0, i.e., $\mathbf{Y}(0) = (0,0,0,0,1,0)^T$. Using (12) and (10) the appropriate final condition to be satisfied is

$$
y_2(1) = 0.
\tag{15}
$$

The eigenvalue R_E is varied until (15) is satisfied to some pre-assigned tolerance; a technique such as the secant method works well to determine R_E. In solving (11) we keep a^2, λ fixed and then find numerically

$$
Ra_E = \max_{\lambda} \min_{a^2} R_E^2(a^2, \lambda),
\tag{16}
$$

varying a^2 for fixed λ and finally maximizing in λ. A very reliable method for determining the minimum and the maximum in (16) is that of golden section search, see e.g., Cheney & Kincaid (1985), p. 462.

For a $2n$th order system the number of ordinary differential equations to be calculated in the compound matrix routine is $^{2n}C_n$. Hence there are 20 equations for a sixth order eigenvalue problem, 70 for an 8th, and so on. For dealing with high order systems an automatic equation generator is obviously desirable. Such a generator is available and further details may be obtained directly from the writer.

References

Adams, R.A. 1975. *Sobolev spaces*. Academic Press, New York.

Ahmadi, G. 1976. Stability of a micropolar fluid layer heated from below. *Int. J. Engng. Sci.* **14**, 81–89.

Alts, T. & Hutter, K. 1988a. Continuum description of the dynamics and thermodynamics of phase boundaries between ice and water. Part I. Surface balance laws and their interpretation in terms of three-dimensional balance laws averaged over the phase change boundary layer. *J. Non-Equilib. Thermodyn.* **13**, 221–257.

Alts, T. & Hutter, K. 1988b. Continuum description of the dynamics and thermodynamics of phase boundaries between ice and water. Part II. Thermodynamics. *J. Non-Equilib. Thermodyn.* **13**, 259–280.

Alts, T. & Hutter, K. 1988c. Continuum description of the dynamics and thermodynamics of phase boundaries between ice and water. Part III. Thermostatics and its consequences. *J. Non-Equilib. Thermodyn.* **13**, 301–329.

Alts, T. & Hutter, K. 1989. Continuum description of the dynamics and thermodynamics of phase boundaries between ice and water. Part IV. On thermostatic stability and well-posedness. *J. Non-Equilib. Thermodyn.* **14**, 1–22.

Andersland, O.B. & Anderson, D.M. 1978. *Geotechnical engineering for cold regions*. McGraw-Hill, New York.

Antar, B.N., Collins, F.G. & Fichtl, G.H. 1980. Influence of solidification on surface tension driven convection. *Int. J. Heat Mass Transfer* **23**, 191-201.

Arnold, V.I. 1965a. Conditions for nonlinear stability of stationary plane curvilinear flows of an ideal fluid. *Dokl. Akad. Nauk SSSR* **162**, 975–978.

Arnold, V.I. 1965b. Variational principle for three-dimensional steady state flows of an ideal fluid. *Prikl. Mat. Mekh.* **29**, 846–851.

Arnold, V.I. 1966a. Sur un principe variationnel pour les écoulements des liquides parfaits et ses applications aux problèmes de stabilité nonlineaires. *J. Mécanique* **5**, 29–43.

Arnold, V.I. 1966b. An a priori estimate in the theory of hydrodynamic stability. *Izv. VUZ Mat.* **54**, 3–5. (English translation 1969, *Amer. Math. Soc. Transl.* **79**.)

Azouni, M.A. 1983. Hysteresis loop in water between 0°C and 4°C. *Geophys. Astrophys. Fluid Dyn.* **24**, 137–142.

Azouni, M.A. & Normand, C. 1983a. Thermoconvective instabilities in a vertical cylinder of water with maximum density effects. I. Experiments. *Geophys. Astrophys. Fluid Dyn.* **23**, 209–222.

Azouni, M.A. & Normand, C. 1983b. Thermoconvective instabilities in a vertical cylinder of water with maximum density effects. II. Theory. *Geophys. Astrophys. Fluid Dyn.* **23**, 223–245.

Bailey, P.B., Chen, P. & Straughan, B. 1984. Stabilization criteria for thermally explosive materials. *Acta Mech.* **53**, 73–79.

Baines, P.G. & Gill, A.E. 1969. On thermohaline convection with linear gradients. *J. Fluid Mech.* **37**, 289–306.

Batchelor, G.K. 1967. *An introduction to fluid dynamics.* Cambridge Univ. Press.

Bénard, H. 1900. Les tourbillons cellulaires dans une nappe liquide. *Revue Gén. Sci. Pure Appl.* **11**, 1261–1271.

Berg, J.C. & Acrivos, A. 1965. The effect of surface active agents on convection cells induced by surface tension. *Chem. Engng. Sci.* **20**, 737–745.

Bernstein, I.B., Frieman, E.A., Kruskal, M.D. & Kulsrud, R.M. 1958. An energy principle for hydromagnetic stability problems. *Proc. Roy. Soc. London A* **244**, 17–40.

Bradley, R. 1978. Overstable electroconvective instabilities. *Q. J. Mech. Appl. Math.* **31**, 381–390.

Briley, P.B., Deemer, A.R. & Slattery, J.C. 1976. Blunt knife-edge and disk surface viscometers. *J. Colloid Interface Sci.* **56**, 1–18.

Busse, F.H. 1967. The stability of finite amplitude cellular convection and its relation to an extremum principle. *J. Fluid Mech.* **30**, 625–649.

Busse, F.H. 1978. Nonlinear properties of thermal convection. *Rep. Prog. Phys.* **41**, 1929–1967.

Busse, F.H. 1988. On the optimum theory of turbulence. In *Energy stability and convection.* Pitman Res. Notes in Math., vol. 168, Longman, Harlow.

Caltagirone, J.P. 1975. Thermoconvective instabilities in a horizontal porous layer. *J. Fluid Mech.* **72**, 269–287.

Caltagirone, J.P. 1980. Stability of a saturated porous layer subject to a sudden rise in surface temperature: comparison between the linear and energy methods. *Q. J. Mech. Appl. Math.* **33**, 47–58.

Carmi, S. 1974. Energy stability of modulated flows. *Phys. Fluids* **17**, 1951–1955.

Castellanos, A., Atten, P. & Velarde, M.G. 1984a. Oscillatory and steady convection in dielectric liquid layers subjected to unipolar injection and temperature gradient. *Phys. Fluids* **27**, 1607–1615.

Castellanos, A., Atten, P. & Velarde, M.G. 1984b. Electrothermal convection: Felici's hydraulic model and the Landau picture of non-equilibrium phase transitions. *J. Non-Equilib. Thermodyn.* **9**, 235–244.

Castillo, J.L. & Velarde, M.G. 1982. Buoyancy-thermocapillary instability: the role of interfacial deformation in one and two-component fluid layers heated from below or above. *J. Fluid Mech.* **125**, 463–474.

Chandra, K. 1938. Instability of fluids heated from below. *Proc. Roy. Soc. London A* **164**, 231–242.

Chandrasekhar, S. 1953. The instability of a layer of fluid heated from below and subject to Coriolis forces. *Proc. Roy. Soc. London A* **217**, 306–327.

Chandrasekhar, S. 1981. *Hydrodynamic and hydromagnetic stability.* Dover, New York.

Chandrasekhar, S. 1987. *Ellipsoidal figures of equilibrium.* Dover, New York.

Chandrasekhar, S. & Elbert, D.D. 1955. The instability of a layer of fluid heated from below and subject to Coriolis forces. II. *Proc. Roy. Soc. London A* **231**, 198–210.

Cheney, W. & Kincaid, D. 1985. *Numerical mathematics and computing.* Brooks-Cole Publishing Co., Monterey, California.

Chhuon, B. & Caltagirone, J.P. 1979. Stability of a horizontal porous layer with timewise periodic boundary conditions. *J. Heat Transfer* **101**, 244–248.

Childress, S., Levandowsky, M. & Spiegel, E.A. 1975. Pattern formation in a suspension of swimming micro-organisms: equations and stability theory. *J. Fluid Mech.* **63**, 591–613.

Chorin, A.J. 1985. Curvature and solidification. *J. Comput. Phys.* **57**, 472–490.

Christopherson, D.G. 1940. Note on the vibation of membranes. *Quart. J. Math.* **11**, 63–65.

Cloot, A. & Lebon, G. 1986. Marangoni convection induced by a nonlinear temperature dependent surface tension. *J. Physique* **47**, 23–29.

Coddington, E.A. & Levinson, N. 1955. *Theory of ordinary differential equations.* McGraw-Hill, New York.

Corte, A.E. 1966. Particle sorting by repeated freezing and thawing. *Biuletyn Peryglacjalny* **15**, 175–240.

Coscia, V. & Padula, M. 1990. Nonlinear energy stability in a compressible atmosphere. *Geophys. Astrophys. Fluid Dyn.* **54**, 49–84.

Cowley, M.D. & Rosensweig, R.E. 1967. The interfacial stability of a ferromagnetic fluid. *J. Fluid Mech.* **30**, 671–688.

Curtis, R.A. 1971. Flows and wave propagation in ferrofluids. *Phys. Fluids* **14**, 2096–2102.

Datta, A.B. & Sastry, V.U.K. 1976. Thermal instability of a horizontal layer of micropolar fluid heated from below. *Int. J. Engng. Sci.* **14**, 631–637.

Davis, S.H. 1969a. On the principle of exchange of stabilities. *Proc. Roy. Soc. London A* **310**, 341–358.

Davis, S.H. 1969b. Buoyancy-surface tension instability by the method of energy. *J. Fluid Mech.* **39**, 347–359.

Davis, S.H. 1971. On the possibility of subcritical instabilities. In *Proc. IUTAM Symp. Herrenalb.* Springer-Verlag, Berlin, Heidelberg, and New York.

Davis, S.H. 1987. Thermocapillary instabilities. *Ann. Rev. Fluid Mech.* **19**, 403–435.

Davis, S.H. & Homsy, G.M. 1980. Energy stability theory for free-surface problems: buoyancy-thermocapillary layers. *J. Fluid Mech.* **98**, 527–553.

Davis, S.H. & von Kerczek, C. 1973. A reformulation of energy stability theory. *Arch. Rational Mech. Anal.* **52**, 112–117.

Davis, S.H., Müller, U. & Dietsche, C. 1984. Pattern selection in single-component systems coupling Bénard convection and solidification. *J. Fluid Mech.* **144**, 133–151.

de Angelis, M. 1990. On universal energy stability of fluid motions in unbounded domains. *Rend. Accad. Sci. Fisiche Matem.* Napoli (Ser. IV) **57**, to appear.

Dennis, J.E. & Schnabel, R.B. 1983. *Numerical methods for unconstrained optimization and nonlinear equations.* Prentice-Hall, Englewood Cliffs.

Deo, B.J.S. & Richardson, A.T. 1983. Generalized energy methods in electrohydrodynamic stability theory. *J. Fluid Mech.* **137**, 131–151.

Drazin, P.G. & Reid, W.H. 1981. *Hydrodynamic stability.* Cambridge Univ. Press.

Dudis, J.J.K. & Davis, S.H. 1971. Energy stability of the buoyancy boundary layer. *J. Fluid Mech.* **47**, 381–403.

Duong, D.D. & Weiland, R.H. 1981. Enzyme deactivation in fixed bed reactors with Michaelis-Menten kinetics. *Biotechnology and Bioengineering* **23**, 691–705.

Embleton, C. & King, C.A.M. 1975. *Glacial and periglacial geomorphology,* 2nd Ed. Wiley, New York.

Eringen, A.C. 1964. Simple microfluids. *Int. J. Engng. Sci.* **2**, 205–217.

Eringen, A.C. 1969. Micropolar fluids with stretch. *Int. J. Engng. Sci.* **7**, 115–127.

Eringen, A.C. 1972. Theory of thermomicrofluids. *J. Math. Anal. Appl.* **38**, 480–496.

Eringen, A.C. 1980. Theory of anisotropic micropolar fluids. *Int. J. Engng. Sci.* **18**, 5–17.

Fearn, D.R., Roberts, P.H. & Soward, A.M. 1988. Convection, stability and the dynamo. In *Energy stability and convection*. Pitman Res. Notes in Math., vol. 168, Longman, Harlow.

Fichera, G. 1978. *Numerical and quantitative analysis*. Pitman Press, London.

Finucane, R.G. & Kelly, R.E. 1976. Onset of instability in a fluid layer heated sinusoidally from below. *Int. J. Heat Mass Transfer* **19**, 71-85.

Finlayson, B.A. 1970. Convective instability of ferromagnetic fluids. *J. Fluid Mech.* **40**, 753–767.

Fosdick, R.L. & Rajagopal, K.R. 1980. Thermodynamics and stability of fluids of third grade. *Proc. Roy. Soc. London A* **339,** 351–377.

Franchi, F. & Straughan, B. 1988. Convection, stability and uniqueness for a fluid of third grade. *Int. J. Nonlinear Mech.* **23**, 377–384.

Gailitis, A. 1977. Formation of the hexagonal pattern on the surface of a ferromagnetic fluid in an applied magnetic field. *J. Fluid Mech.* **82**, 401–413.

Galdi, G.P. 1985. Nonlinear stability of the magnetic Bénard problem via a generalized energy method. *Arch. Rational Mech. Anal.* **87**, 167–186.

Galdi, G.P. & Padula, M. 1990. A new approach to energy theory in the stability of fluid motion. *Arch. Rational Mech. Anal.* **110,** 187–286.

Galdi, G.P., Payne, L.E., Proctor, M.R.E. & Straughan, B. 1987. Convection in thawing subsea permafrost. *Proc. Roy. Soc. London A* **414**, 83–102.

Galdi, G.P. & Rionero, S. 1985. *Weighted energy methods in fluid dynamics and elasticity*. Lecture Notes in Math., vol. 1134, Springer-Verlag, New York.

Galdi, G.P. & Straughan, B. 1985a. Exchange of stabilities, symmetry and nonlinear stability. *Arch. Rational Mech. Anal.* **89**, 211–228.

Galdi, G.P. & Straughan, B. 1985b. A nonlinear analysis of the stabilizing effect of rotation in the Bénard problem. *Proc. Roy. Soc. London A* **402**, 257–283.

Galdi, G.P. & Straughan, B. 1987. A modified model problem of Drazin and Reid exhibiting sharp conditional stability. *Ann. Univ. Ferrara* **32**, 39–43.

Galdi, G.P. & Straughan, B. 1990. A nonlinear analysis of electrohydrodynamic stability. *To appear.*

George, J.H., Gunn, R.D. & Straughan, B. 1989. Patterned ground formation and penetrative convection in porous media. *Geophys. Astrophys. Fluid Dyn.* **46**, 135–158.

Gilbarg, D. & Trudinger, N.S. 1977. *Elliptic partial differential equations of second order.* Springer-Verlag, Berlin, Heidelberg, and New York.

Gill, A.E. 1966. The boundary-layer regime for convection in a rectangular cavity. *J. Fluid Mech.* **26**, 515–536.

Gill, A.E. 1969. A proof that convection in a porous vertical slab is stable. *J. Fluid Mech.* **35**, 545–547.

Giorgi, F. 1989. Two-dimensional simulations of possible mesoscale effects of nuclear war fires. I. Model description. *J. Geophys. Res. D* **94**, 1127–1144.

Gleason, K.J. 1984. *Nonlinear Boussinesq convection in porous media: application to patterned ground formation.* M.S. Thesis, University of Colorado, Boulder.

Gleason, K.J., Krantz, W.B., Caine, N., George, J.H. & Gunn, R.D. 1986. Geometrical aspects of sorted patterned ground in recurrently frozen soil. *Science* **232**, 216–220.

Goldthwaite, R.P. 1976. Frost sorted patterned ground: a review. *Quarternary Research* **6**, 27–35.

Gresho, P.M. & Sani, R.L. 1970. The effects of gravity modulation on the stability of a heated fluid layer. *J. Fluid Mech.* **40**, 783–806.

Gripp, K. 1926. Uber frost und strukturboden auf Spitzbergen. *Gesell. Erdkunde Berlin Zeitschr.* 351–354.

Gross, M.J. 1967. Article in *Mantles of the Earth and terrestial planets.* Wiley, New York.

Gumerman, R.J. & Homsy, G.M. 1975. The stability of uniformly accelerated flows with application to convection driven by surface tension. *J. Fluid Mech.* **68**, 191–207.

Gurtin, M.E. 1971. On the thermodynamics of chemically reacting mixtures. *Arch. Rational Mech. Anal.* **43**, 198-212.

Hardy, G.H., Littlewood, J.E. & Polya, G. 1934. *Inequalities.* Cambridge Univ. Press.

Harrison, W.D. 1982. Formulation of a model for pore water convection in thawing subsea permafrost. *Mitteilungen der Versuchsanstalt für Wasserbau, Hydrologie und Glaziologie.* vol. 57, ETH, Zürich.

Harrison, W.D. & Osterkamp, T.E. 1982. Measurements of the electrical conductivity of interstitial water in subsea permafrost. In *Proc. 4th Canadian Permafrost Conf.* National Research Council, Canada.

Hills, R.N. & Roberts, P.H. 1987a. Relaxation effects in a mixed phase region. I. General theory. *J. Non-Equilib. Thermodyn.* **12**, 169–181.

Hills, R.N. & Roberts, P.H. 1987b. Relaxation effects in a mixed phase region. II. Illustrative examples. *J. Non-Equilib. Thermodyn.* **12**, 183–195.

Hills, R.N. & Roberts, P.H. 1988a. On the formulation of diffusive mixture theories for two-phase regions. *J. Engng. Math.* **22**, 93–106.

Hills, R.N. & Roberts, P.H. 1988b. A generalized Scheil-Pfann equation for a dynamical theory of a mushy zone. *Int. J. Non-Linear Mechs.* **23**, 327–339.

Hills, R.N. & Roberts, P.H. 1990. On the motion of a fluid that is incompressible in a generalized sense and its relationship to the Boussinesq approximation. *Stab. Appl. Anal. Cont. Media*, to appear.

Holm, D.D., Marsden, J.E., Ratiu, T. & Weinstein, A. 1985. Nonlinear stability of fluid and plasma equilibria. *Physics Reports* **123**, 1–116.

Homsy, G.M. 1973. Global stability of time-dependent flows: impulsively heated or cooled fluid layers. *J. Fluid Mech.* **60**, 129–139.

Homsy, G.M. 1974. Global stability of time-dependent flows. Part 2. Modulated fluid layers. *J. Fluid Mech.* **62**, 387–403.

Hurle, D.T.J. & Jakeman, E. 1971. Soret-driven thermosolutal convection. *J. Fluid Mech.* **47**, 667–687.

Hurle, D.T.J., Jakeman, E. & Pike, E.R. 1967. On the solution of the Bénard problem with boundaries of finite conductivity. *Proc. Roy. Soc. London A* **296**, 469-475.

Hurle, D.T.J., Jakeman, E. & Wheeler, A.A. 1982. Effect of solutal convection on the morphological stability of a binary alloy. *J. Crystal Growth* **58**, 163–179.

Hutter, K. 1982. A mathematical model of polythermal glaciers and ice sheets. *Geophys. Astrophys. Fluid Dyn.* **21**, 201–224.

Hutter, K. 1983. *Theoretical Glaciology.* D. Reidel, Dordrecht, Boston, and Lancaster.

Hutter, K., Blatter, H. & Funk, M. 1988. A model computation of moisture content in polythermal glaciers. *J. Geophys. Res.* **93**, 12205-12214.

Hutter, K. & Engelhardt, H. 1988. The use of continuum thermodynamics in the formulation of ice-sheet dynamics. *Annals Glaciol.* **11**, 46–51.

Hutter, K. & Straughan, B. 1984. A scaling analysis for convection in thawing subsea permafrost. *In preparation.*

Ivers, D.J. & James, R.W. 1984. Axisymmetric anti-dynamo theorems in compressible non-uniform conducting fluids. *Phil. Trans. Roy. Soc. London A* **312**, 179–218.

Ivers, D.J. & James, R.W. 1986. Extension of the Namikawa-Matsushita antidynamo theorem to toroidal fields. *Geophys. Astrophys. Fluid Dyn.* **36**, 317–324.

Jankowski, D.F., Neitzel, G.P., Shen, Y.H. & Squire, T.H. 1988. Stability of thermocapillary convection during float-zone crystal growth. In *Energy stability and convection*. Pitman Res. Notes in Math., vol. 168, Longman, Harlow.

Jhaveri, B.S. & Homsy, G.M. 1982. The onset of convection in fluid layers heated rapidly in a time-dependent manner. *J. Fluid Mech.* **114**, 251–260.

Joseph, D.D. 1965. On the stability of the Boussinesq equations. *Arch. Rational Mech. Anal.* **20**, 59–71.

Joseph, D.D. 1966. Nonlinear stability of the Boussinesq equations by the method of energy. *Arch. Rational Mech. Anal.* **22**, 163–184.

Joseph, D.D. 1970. Global stability of the conduction-diffusion solution. *Arch. Rational Mech. Anal.* **36**, 285–292.

Joseph, D.D. 1976a. Stability of fluid motions. Vol. I. Springer-Verlag, Berlin, Heidelberg, and New York.

Joseph, D.D. 1976b. Stability of fluid motions. Vol. II. Springer-Verlag, Berlin, Heidelberg, and New York.

Joseph, D.D. 1988. Two fluids heated from below. In *Energy stability and convection*. Pitman Res. Notes in Math., vol. 168, Longman, Harlow.

Joseph, D.D. & Hung, W. 1971. Contributions to the nonlinear theory of stability of viscous flow in pipes and between rotating cylinders. *Arch. Rational Mech. Anal.* **44**, 1–22.

Joseph, D.D. & Shir, C.C. 1966. Subcritical convective instability. Part 1. Fluid layers. *J. Fluid Mech.* **26**, 753–768.

Jou, D., Casas-Vázquez, J. & Lebon, G. 1988. Extended irreversible thermodynamics. *Rep. Prog. Phys.* **51**, 1105–1179.

Kassoy, D.R. 1980. A guide to mathematical models of convection processes in geothermal systems. In *Fluid mechanics in energy conversion*. SIAM, Philadelphia.

Kato, T. 1976. *Perturbation theory for linear operators*. Springer-Verlag, Berlin, Heidelberg, and New York.

Kaye, J.A. & Rood, R.B. 1989. Chemistry and transport in a three-dimensional stratospheric model: chlorine species during a simulated stratospheric warming. *J. Geophys. Res. D* **94**, 1057–1083.

Kelvin, Lord. 1887. On the stability of steady and of periodic fluid motion. *Phil. Mag. (5)* **23**, 459–464; 529–539.

Kerr, O.S. 1989. Heating a salinity gradient from a vertical sidewall: linear theory. *J. Fluid Mech.* **207**, 323–352.

Kerr, O.S. 1990. Heating a salinity gradient from a vertical sidewall: nonlinear theory. *J. Fluid Mech.* **217**, 529–546.

Kloeden, P. & Wells, R. 1983. An explicit example of Hopf bifurcation in fluid mechanics. *Proc. Roy. Soc. London A* **390**, 293–320.

Kondo, M.A. & Unno, W. 1982. Convection and gravity waves in two layer models. I. Overstable modes driven in conducting boundary layers. *Geophys. Astrophys. Fluid Dyn.* **22**, 305–324.

Kondo, M.A. & Unno, W. 1983. Convection and gravity waves in two layer models. II. Overstable convection of large horizontal scales. *Geophys. Astrophys. Fluid Dyn.* **27**, 229–252.

Lalas, D.P. & Carmi, S. 1971. Thermoconvective stability of ferrofluids. *Phys. Fluids* **14**, 436–437.

Landau, L.D. & Lifshitz, E.M. 1959. *Fluid Mechanics.* Pergamon Press, London.

Landau, L.D., Lifshitz, E.M. & Pitaevskii, L.P. 1984. *Electrodynamics of continuous media.* (2nd. Edition), Pergamon Press, Oxford.

Langlois, W.E. 1985. Buoyancy-driven flows in crystal growth melts. *Ann. Rev. Fluid Mech.* **17**, 191-215.

Lebon, G. & Perez-Garcia, C. 1981. Convective instability of a micropolar fluid layer by the method of energy. *Int. J. Engng. Sci.* **19**, 1321–1329.

Levandowsky, M., Childress, S., Spiegel, E.A. & Hutner, S.H. 1975. A mathematical model of pattern formation by swimming micro-organisms. *J. Protozoology* **22**, 296–306.

Levine, H.A. 1973. Some nonexistence and instability theorems for solutions of formally parabolic equations of the form $Pu_t = -Au + F(u)$. *Arch. Rational Mech. Anal.* **51**, 371–386.

Levine, H.A., Payne, L.E., Sacks, P.E. & Straughan, B. 1989. Analysis of a convective reaction-diffusion equation. *SIAM J. Math. Anal.* **20**, 133–147.

Lindsay, K.A. & Straughan, B. 1979. A thermodynamic viscous interface theory and associated stability problems. *Arch. Rational Mech. Anal.* **71**, 307–326.

Lindsay, K.A. & Straughan, B. 1990. Energy methods for nonlinear stability in convection problems, primarily related to geophysics. *Continuum Mech. Thermodyn.* **2**, 245–277.

Loper, D.E. & Roberts, P.H. 1978. On the motion of an Iron-Alloy core containing a slurry. I. General theory. *Geophys. Astrophys. Fluid Dyn.* **9**, 289–321.

Loper, D.E. & Roberts, P.H. 1980. On the motion of an Iron-Alloy core containing a slurry. II. A simple model. *Geophys. Astrophys. Fluid Dyn.* **16**, 83–127.

Loper, D.E. & Roberts, P.H. 1981. A study of conditions at the inner core boundary of the Earth. *Phys. Earth Planetary Interiors* **24**, 302–307.

Lortz, D. & Meyer-Spasche, R. 1982. On the decay of symmetric dynamo fields. *Math. Meth. Appl. Sci.* **4**, 91–97.

Low, A.R. 1925. Instability of viscous fluid motion. *Nature* **229**A, 299–300.

Lutyen, P.J. 1987. Stability of a rotating liquid drop immersed in a corotating fluid with different density. *Proc. Roy. Soc. London A* **414**, 59–82.

McFadden, G.B., Rehm, R.G., Coriell, S.R., Chuck, W. & Morrish, K.A. 1984. Thermosolutal convection during directional solidification. *Metall. Trans.* **15**A, 2125–2137.

McKay, G. & Straughan, B. 1990. Nonlinear energy stability and convection near the density maximum. *Acta Mech.* To appear.

McTaggart, C.L. 1983a. *Energy and linear stability analyses of surface and second sound effects in convection problems.* Ph.D. Thesis. University of Glasgow.

McTaggart, C.L. 1983b. Convection driven by concentration and temperature dependent surface tension. *J. Fluid Mech.* **134**, 301–310.

McTaggart, C.L. 1984. On the stabilizing effect of surface films in Bénard convection. *Physico Chemical Hydrodyn.* **5**, 321–331.

Malikkides, C.O. & Weiland, R.H. 1982. On the mechanism of immobilized glucose oxidase deactivation by hydrogen peroxide. *Biotechnology and Bioengineering* **24**, 2419–2439.

Maremonti, P. 1988. On the asymptotic behaviour of the L^2-norm of suitable weak solutions to the Navier-Stokes equations in three-dimensional exterior domains. *Commun. Math. Phys.* **118**, 385–400.

Martin, P.J. & Richardson, A.T. 1984. Conductivity models of electrothermal convection in a plane layer of dielectric liquid. *J. Heat Transfer* **106**, 131–136.

Merker, G.P., Waas, P. & Grigull, U. 1979. Onset of convection in a horizontal water layer with maximum density effects. *Int. J. Heat Mass Transfer* **22**, 505–515.

Morland, L.W., Kelly, R.J. & Morris, E.M. 1990. A mixture theory for a phase-changing snowpack. *Cold Regions Sci. Tech.* To appear.

Morro, A. & Romeo, M. 1988. The law of mass action for fluid mixtures with several temperatures and velocities. *J. Non-Equilib. Thermodyn.* **13**, 339-353.

Morro, A. & Straughan, B. 1990. Convective instabilities for reacting viscous fluids far from equilibrium. *J. Non-Equilib. Thermodyn.* **15**, 139-150.

Müller, I. 1985. *Thermodynamics.* Pitman, Boston.

Müller, I. & Ruggeri, T. (eds.) 1987. *Kinetic theory and extended thermodynamics.* Pitagora, Bologna.

Müller-Beck, H. 1966. Paleohunters in America: origins and diffusion. *Science* **152**, 1191-1210.

Mulone, G. 1990. On the stability of plane parallel convective flow. *Acta Mech.* To appear.

Mulone, G. & Rionero, S. 1989. On the nonlinear stability of the rotating Bénard problem via the Lyapunov direct method. *J. Math. Anal. Appl.* **144**, 109–127.

Munson, B.R. & Joseph, D.D. 1971. Viscous incompressible flow between concentric rotating cylinders. Part 2. Hydrodynamic stability. *J. Fluid Mech.* **49**, 305–318.

Neitzel, G.P. 1982. Onset of convection in impulsively heated or cooled fluid layers. *Phys. Fluids* **25**, 210–211.

Niedrauer, T.M. & Martin, S. 1979. An experimental study of brine drainage and convection in young sea ice. *J. Geophys. Res.* C **84**, 1176–1186.

Nordenskjold, O. 1909. *Die Polarwelt*. B.G. Teubner, Berlin.

Osterkamp, T.E. & Harrison, W.D. 1982. Temperature measurements in subsea permafrost off the coast of Alaska. *Proc. 4th Canadian Permafrost Conf.* National Research Council, Canada.

Padula, M. 1986. Nonlinear energy stability for the compressible Bénard problem. *Boll. U.M.I.* **5**-B, 581–602.

Padula, M. 1988. Energy instability methods: an application to the Burger equation. In *Energy stability and convection*. Pitman Res. Notes in Math., vol. 168, Longman, Harlow.

Pandiz, S.N. & Seinfeld, J.H. 1989. Sensitivity analysis of a chemical mechanism for aqueous-phase atmospheric chemistry. *J. Geophys. Res.* D **94**, 1105–1126.

Payne, L.E., Song, J.C. & Straughan, B. 1988. Double diffusive porous penetrative convection; thawing subsea permafrost. *Int. J. Engng. Sci.* **26**, 797–809.

Payne, L.E. & Straughan, B. 1987. Unconditional nonlinear stability in penetrative convection. *Geophys. Astrophys. Fluid Dyn.* **39**, 57–63. (Also, Corrected and extended numerical results. *Geophys. Astrophys. Fluid Dyn.* (1988) **43**, 307–309.)

Payne, L.E. & Straughan, B. 1989. Critical Rayleigh numbers for oscillatory and nonlinear convection in an isotropic thermomicropolar fluid. *Int. J. Engng. Sci.* **27**, 827–836.

Pearlstein, A.J., Harris, R.M. & Terrones, G. 1989. The onset of convective instability in a triply diffusive fluid layer. *J. Fluid Mech.* **202,** 443–465.

Proctor, M.R.E. 1981. Steady subcritical thermohaline convection. *J. Fluid Mech.* **105**, 507–521.

Ray, R.J., Krantz, W.B., Caine, T.N. & Gunn, R.D. 1983. A model for sorted patterned-ground regularity. *J. Glaciology* **29**, 317–337.

Reddy, B.D. & Voyé, H.F. 1988. Finite element analysis of the stability of fluid motions. *J. Comput. Phys.* **79**, 92–112.

Richardson, A.T. 1988. A hydraulic model of electrothermal convection in a plane layer of dielectric liquid. *Physico Chem. Hydrodyn.* **10**, 355–367.

Rionero, S. 1967. Sulla stabilità asintotica in media in magnetoidrodinamica. *Ann. Matem. Pura Appl.* **76**, 75–92.

Rionero, S. 1968. Metodi variazionali per la stabilità asintotica in media in magnetoidrodinamica. *Ann. Matem. Pura Appl.* **78**, 339–364.

Rionero, S. 1971. Sulla stabilità magnetofluidodinamica nonlineare asintotica in media in presenza di effetto Hall. *Ricerche Matem.* **20**, 285–296.

Rionero, S. 1988. On the choice of the Lyapunov function in the stability of fluid motions. In *Energy stability and convection*. Pitman Res. Notes in Math., vol. 168, Longman, Harlow.

Rionero, S. & Mulone, G. 1987. On the non-linear stability of a thermodiffusive fluid mixture in a mixed problem. *J. Math. Anal. Appl.* **124**, 165–188.

Rionero, S. & Mulone, G. 1988. A nonlinear stability analysis of the magnetic Bénard problem through the Lyapunov direct method. *Arch. Rational. Mech. Anal.* **103**, 347–368.

Rionero, S. & Straughan, B. 1990. Convection in a porous medium with internal heat source and variable gravity effects. *Int. J. Engng. Sci.* **28**, 497–503.

Roberts, P.H. 1967. *An introduction to magnetohydrodynamics.* Longman, London.

Roberts, P.H. 1969. Electrohydrodynamic convection. *Q. J. Mech. Appl. Math.* **22**, 211–220.

Roberts, P.H. 1981. Equilibria and stability of a fluid type II superconductor. *Q. J. Mech. Appl. Math.* **34**, 327–343.

Roberts, P.H. 1987a. Convection in spherical systems. In *Irreversible phenomena and dynamical systems analysis in geosciences* (Eds. C. Nicolis & G. Nicolis). D. Reidel, Dordrecht, Boston, and Lancaster, pp. 53–71.

Roberts, P.H. 1987b. Dynamo theory. In *Irreversible phenomena and dynamical systems analysis in geosciences* (Eds. C. Nicolis & G. Nicolis). D. Reidel, Dordrecht, Boston, and Lancaster, pp. 73–133.

Roberts, P.H. 1988a. Future of geodynamo theory. *Geophys. Astrophys. Fluid Dyn.* **44**, 3–31.

Roberts, P.H. 1988b. On topographic core-mantle coupling. *Geophys. Astrophys. Fluid Dyn.* **44**, 181–187.

Roberts, P.H. & Stewartson, K. 1974. On finite amplitude convection in a rotating magnetic system. *Phil. Trans. Roy. Soc. London A* **277**, 287–315.

Rodriguez-Luis, A., Castellanos, A. & Richardson, A.T. 1986. Stationary instabilities in a dielectric liquid layer subjected to an arbitrary unipolar injection and an adverse thermal gradient. *J. Phys. D: Appl. Phys.* **19**, 2115–2122.

Rood, R.B. 1987. Numerical advection algorithms and their role in atmospheric transport and chemistry models. *Reviews Geophys.* **25**, 71–100.

Rosenblat, S., Davis, S.H. & Homsy, G.M. 1982. Nonlinear Marangoni convection in bounded layers. I. Circular cylindrical containers. *J. Fluid Mech.* **120**, 91–122.

Rosenblat, S., Homsy, G.M. & Davis, S.H. 1982. Nonlinear Marangoni convection in bounded layers. II. Rectangular cylindrical containers. *J. Fluid Mech.* **120**, 123–138.

Rosenblat, S. & Tanaka, G.A. 1971. Modulation of thermal convection instability. *Phys. Fluids* **14**, 1319–1322.

Rosensweig, R.E. 1985. *Ferrohydrodynamics.* Cambridge Univ. Press.

Rossby, H.T. 1969. A study of Bénard convection with and without rotation. *J. Fluid Mech.* **36**, 309–335.

Ruddick, B.R. & Shirtcliffe, T.G.L. 1979. Data for double diffusers: physical properties of aqueous salt-sugar solutions. *Deep Sea Research* **26**, 775-787.

Scriven, L.E. 1960. Dynamics of a fluid interface. Equation of motion for Newtonian surface fluids. *Chem. Engng. Sci.* **12**, 98–108.

Segel, L.A. & Jackson, J.L. 1972. Dissipative structure: an explanation and an ecological example. *J. Theoretical Biol.* **37**, 545–559.

Serrin, J. 1959a. Mathematical principles of classical fluid mechanics. In *Handbuch der Physik*, vol. III/1. Springer-Verlag, Berlin, Göttingen, and Heidelberg.

Serrin, J. 1959b. On the stability of viscous fluid motions. *Arch. Rational Mech. Anal.* **3**, 1–13.

Shir, C.C. & Joseph, D.D. 1968. Convective instability in a temperature and concentration field. *Arch. Rational Mech. Anal.* **30**, 38–80.

Shliomis, M.I. 1974. Magnetic fluids. *Sov. Phys. Usp.* **17**, 153–169.

Simo, J.C., Posbergh, T.A. & Marsden, J.E. 1990. Stability of coupled rigid body and geometrically exact rods: block diagonalization and the energy momentum method. *Physics Reports* **193**, 279–360.

Slemrod, M. 1978. An energy stability method for simple fluids. *Arch. Rational Mech. Anal.* **68**, 1–18.

Spiegel, E.A. 1965. Convective instabilities in a compressible atmosphere. I. *Astrophys. J.* **141**, 1068–1090.

Straughan, B. 1982. *Instability, nonexistence and weighted energy methods in fluid dynamics and related theories.* Pitman Press, London.

Straughan, B. 1983. Energy stability in the Bénard problem for a fluid of second grade. *ZAMP* **34**, 502–509.

Straughan, B. 1985. Finite amplitude instability thresholds in penetrative convection. *Geophys. Astrophys. Fluid Dyn.* **34**, 227–242.

Straughan, B. 1987. Stability of a layer of dipolar fluid heated from below. *Math. Meth. Appl. Sci.* **9**, 35–45.

Straughan, B. 1988. A nonlinear analysis of convection in a vertical porous slab. *Geophys. Astrophys. Fluid Dyn.* **42**, 269–275.

Straughan, B. 1989. Convection in a variable gravity field. *J. Math. Anal. Appl.* **140**, 467–475.

Straughan, B. 1990. A nonlinear energy analysis for thermo-convective ferrohydrodynamic stability. *To appear.*

Straughan, B. 1991. A note on surface film driven convection. *Glasgow Math. J.* **33**, 155–158.

Straughan, B., Bampi, F. & Morro, A. 1984. Chemical convective instability and quasi-equilibrium thermodynamics. *Meccanica* **19**, 291–293.

Straughan, B., Ewing, R.E., Jacobs, P.G. & Djomehri, M.J. 1987. Nonlinear instability for a modified form of Burgers' equation. *Numer. Meth. Part. Diff. Equns.* **3**, 51–64.

Stredulinsky, E.W., Meyer-Spasche, R. & Lortz, D. 1986. Asymptotic behaviour of solutions of certain parabolic problems with space and time dependent coefficients. *Comm. Pure Appl. Math.* **39**, 233–266.

Swift, D.W. & Harrison, W.D. 1984. Convective transport of brine and thaw of subsea permafrost: results of numerical simulations. *J. Geophys. Res.* *C2* **89**, 2080–2086.

Swift, D.W., Harrison, W.D. & Osterkamp, T.E. 1982. Heat and salt transport processes in thawing subsea permafrost at Prudhoe Bay, Alaska. In *Proc. 4th Intl. Permafrost Conf.* National Academy Press, Washington, D.C.

Takashima, M. & Aldridge, K.D. 1976. The stability of a horizontal layer of dielectric fluid under the simultaneous action of a vertical DC electric field and a vertical temperature gradient. *Q. J. Mech. Appl. Math.* **29**, 71–87.

Temam, R. 1978. *The Navier-Stokes equations.* North-Holland, Amsterdam.

Terrones, G. & Pearlstein, A.J. 1989. The onset of convection in a multicomponent fluid layer. *Phys. Fluids A* **1**, 845–853.

Thomson, J. 1882. On a changing tesselated structure in certain liquids. *Proc. Phil. Soc. Glasgow* **13**, 464–468.

Turnbull, R.J. 1968a. Electroconvective instability with a stabilizing temperature gradient. I. Theory. *Phys. Fluids* **11**, 2588–2596.

Turnbull, R.J. 1968b. Electroconvective instability with a stabilizing temperature gradient. II. Experimental results. *Phys. Fluids* **11**, 2597–2603.

Veronis, G. 1959. Cellular convection with finite amplitude in a rotating fluid. *J. Fluid Mech.* **5**, 401–435.

Veronis, G. 1963. Penetrative convection. *Astrophys. J.* **137**, 641–663.

Veronis, G. 1965. On finite amplitude instability in thermohaline convection. *J. Marine Res.* **23**, 1–17.

Veronis, G. 1966. Motions at subcritical values of the Rayleigh number in a rotating fluid. *J. Fluid Mech.* **24**, 545–554.

Veronis, G. 1968a. Effect of a stabilizing gradient of solute on thermal convection. *J. Fluid Mech.* **34**, 315–336.

Veronis, G. 1968b. Large amplitude Bénard convection in a rotating fluid. *J. Fluid Mech.* **31**, 113–139.

Viola, T. 1941. Calcolo approssimato di autovalori. *Rend. Matematica* Ser. V **2**, 71–106.

Walden, R.W. & Ahlers, G. 1981. Non-Boussinesq and penetrative convection in a cylindrical cell. *J. Fluid Mech.* **109**, 89-114.

Washburn, A.L. 1973. *Periglacial processes and environments*. St. Martin's Press, New York.

Washburn, A.L. 1980. *Geocryology*. John Wiley & Sons, New York.

Whitehead, J.A. 1971. Upon boundary conditions imposed by a stratified fluid. *Geophys. Fluid Dyn.* **2**, 289–298.

Whitehead, J.A. & Chen, M.M. 1970. Thermal instability and convection of a thin layer bounded by a stably stratified region. *J. Fluid Mech.* **40**, 549–576.

Wollkind, D.J. & Bdzil, J. 1971. Comments on chemical instabilities. *Phys. Fluids* **14**, 1813–1814.

Wollkind, D.J. & Frisch, H.L. 1971. Chemical instabilities: I. A heated horizontal layer of dissociating fluids. *Phys. Fluids* **14**, 13–18.

Worraker, W.J. & Richardson, A.T. 1979. The effect of temperature-induced variations in charge carrier mobility on a stationary electrohydrodynamnic instability. *J. Fluid Mech.* **93**, 29–45.

Ybarra, P.L.G. & Velarde, M.G. 1979. The influence of the Dufour and Soret effects on the stability of a binary gas layer heated from below or above. *Geophys. Astrophys. Fluid Dyn.* **13**, 83–94.

Yih, C.S. & Li, C.H. 1972. Instability of unsteady flows or configurations. Part 2. Convective instability. *J. Fluid Mech.* **54**, 143–152.

Zahn, J.P., Toomre, J. & Latour, J. 1982. Nonlinear modal analysis of penetrative convection. *Geophys. Astrophys. Fluid Dyn.* **22**, 159–193.

Index

Applied Mathematical Sciences

cont. from page ii